Sanitation for All: A Women's Perspective

Sanitation for All: A Women's Perspective

Blanca Elena Jiménez Cisneros,
Akissa Bahri, Juliana Calabria de Araújo,
Claudia Campos, María Luisa Castro de
Esparza, Bettina Genthe, Paula Kehoe,
Corinne Schuster-Wallace, Kwanrawee
Sirikanchana and Maria Inês Zanoli Sato

Published by   **IWA Publishing**
Unit 104–105, Export Building
1 Clove Crescent
London E14 2BA, UK
Telephone: +44 (0)20 7654 5500
Fax: +44 (0)20 7654 5555
Email: publications@iwap.co.uk
Web: www.iwapublishing.com

First published 2024
© 2024 IWA Publishing

The publisher makes no representation, express or implied, with regard to the accuracy of the information contained in this book and cannot accept any legal responsibility or liability for errors or omissions that may be made.

Disclaimer
The information provided and the opinions given in this publication are not necessarily those of IWA and should not be acted upon without independent consideration and professional advice. IWA and the Editors and Authors will not accept responsibility for any loss or damage suffered by any person acting or refraining from acting upon any material contained in this publication.

British Library Cataloguing in Publication Data
A CIP catalogue record for this book is available from the British Library

ISBN: 9781789064032 (paperback)
ISBN: 9781789064049 (eBook)
ISBN: 9781789064056 (ePub)

Doi: 10.2166/9781789064049

This eBook was made Open Access in August 2024.

Contents

The authors wish to thank the Bill and Melinda Gates Foundation, notably Dr Doulaye Kone, Interim Director, Water, Sanitation and Hygiene, for their support in enabling open access publication of this book.

The Authors

Blanca Elena Jiménez Cisneros is an Environmental Engineer with a PhD in wastewater treatment. Presently, she is the Ambassador of Mexico to France and a Senior Researcher at the Engineering Institute of Universidad Nacional Autónoma de México (UNAM). She was the Director of the Water Science Division at UNESCO and the General Director of the National Water Commission of Mexico. She is a former member of Water Reuse, Disinfection and Biosolids IWA Specialist Groups.

Akissa Bahri holds a PhD in Water Resources Engineering. She served as Minister of Agriculture, Water Resources and Fisheries in Tunisia. She has worked in Water Resources Management and Water Reuse in various capacities as a Professor and Director of Research in Tunisia, as well as Coordinator of the African Water Facility at the African Development Bank, and Director for Africa at the International Water Management Institute. She is also part of the IWA Water Reuse Specialist Group.

Juliana Calábria de Araújo is an Associate Professor in wastewater microbiology at the Federal University of Minas Gerais, Brazil. Associate-Editor of the Cleaner Water journal, she coordinated several national projects with international partners, including the SARS-CoV-2 sewage surveillance in Belo Horizonte. Currently, she coordinates the Genomic Sewage Surveillance Project, and is a member of the Climade (Climate amplified Diseases & Epidemics) Consortium and is a CNPq Research Productivity Fellow.

Claudia Campos holds a PhD in Environmental Microbiology. Formerly a professor at the Pontificia Universidad Javeriana in Bogotá, Colombia, she specializes in topics related to Water and Biosolids Reuse for Agriculture, and the evaluation of toxicity in water. She has signed over 60 publications in scientific journals and has worked with the United Nations Development Program and the Ministry of Science and Technology in Colombia.

María Luisa Castro de Esparza is a chemist and holds a degree in Sanitary and Environmental Engineering, a master's in Environmental Management and a PhD in Public Health. She has worked for 40 years at the Pan American Center for Sanitary Engineering and Environmental Sciences (CEPIS/PAHO/WHO) as a regional advisor providing technical advice on water quality, environmental health, to laboratories in Latin America and the Caribbean. She is also a consultant for the World Bank, Japan International Cooperation Agency (JICA), International Development Research Center (IDRC) and US Environmental Protection Agency (USEPA).

Bettina Genthe is a microbiologist and senior researcher with over 40 years of experience in environmental health aspects of water quality and health risk assessments. She is a member of an independent international panel for wastewater reuse in South Africa. She has been a temporary advisor to the World Health Organization and USEPA. She has extensive experience in risk-based water quality guideline development.

Paula Kehoe is the Director of Water Resources within the San Francisco Public Utilities Commission and is responsible for diversifying San Francisco's water supply portfolio through the implementation of water conservation, groundwater, recycled water, onsite water recycling, innovations, and purified water programs. Paula manages the OneWaterSF program, serves as the Chair of the National Blue Ribbon Commission for Onsite Water Systems, and is a Board Member of the WateReuse Association.

Corinne Schuster-Wallace is the Executive Director of the Global Institute for Water Security and faculty member in the Department of Geography and Planning, University of Saskatchewan, Canada. Previous positions include Senior Research Fellow for UNU Institute for Water, Environment, and Public Health Agency of Canada. She founded the Women Plus Water Community and is the inaugural Chair of Cansu Global for the Science, Innovation, Technology, and Education initiative.

Kwanrawee Sirikanchana is a Senior Research Scientist at the Chulabhorn Research Institute, Thailand. Her areas of expertise include water pollution control, water and health, disinfection, antimicrobial resistance and quantitative microbial risk assessment. She is part of the National Committee for setting water quality standards from wastewater effluent sources and is a Board member of the Health-Related Water Microbiology IWA Specialist group.

Maria Inês Zanoli Sato is a biomedical scientist with a PhD in Environmental Microbiology. She has managed the Environmental Analysis Department of the São Paulo State Environmental Company (CETESB), Brazil, for over 20 years, leading a laboratory complex responsible for the analytical and research support to the environmental quality monitoring programs and inspection and licensing activities. She is a Technical Advisor to the Brazilian Ministry of Health, Pan American Health Organization (PAHO)/World Health Organization (WHO)/United Nations Environment Programme (UNEP) and a member of the IWA Health-Related Water Microbiology Specialist Group.

Acronyms and abbreviations

AfBD	African Development Bank
AIT	Asian Institute of Technology
ANA	National Water and Basic Sanitation Agency of Brazil
AUA	Asociación de usuarios de agua
BNBC	Bangladesh National Building Code
BNH	National Housing Bank
BOD	biochemical oxygen demand
BOT	build, operate and transfer contract
CESB	state basic sanitation companies, in Portuguese
CHP	combined heat and power
CONAGUA	Comisión Nacional del Agua
COVID-19	coronavirus disease 2019
CSR	corporate social responsibility
CWIS	citywide inclusive sanitation
DBO	design, build and operate contract
EFSA	European Food Safety Authority
EPA	Environmental Protection Agency
ESA	external support agency
ESAWAS	Eastern and Southern Africa Water and Sanitation Association
FAT	worker support fund
FGTS	severance indemnity fund / Fundo de Garantia do Tempo de Servicio [in Portuguese]
FSM	faecal sludge management
GDP	gross domestic product
GHG	greenhouse gas
GLAAS	Global Analysis and Assessment of Sanitation and Drinking Water
GWP	global water partnership
H_2S	hydrogen sulphide
HRtWS	human rights to water and sanitation

JAC	Junta de Acción Comunal [in Spanish]
JMP WHO/UNICEF	Joint Monitoring Programme for water supply, sanitation and hygiene
KCC	Khulna City Corporation
KWASA	Khulna Water Supply and Sewerage Authority
LPG	liquefied petroleum gas
MoLGRD&C	Ministry of Local Government, Rural Development and Cooperatives
$MTCO_2e$	metric tonnes of carbon dioxide equivalent
N	nitrogen
n	sample size
NBRC	National Blue-Ribbon Commission
NBS	nature-based solution
NGO	non-governmental organisation
NSS	non-sewered system
ODA	official development assistance
OECD	Organisation for Economic Co-operation and Development
O&M	operation and maintenance
OSS	on-site sanitation system
P	phosphorus
PAC	Programa de Aceleração de Crescimento [in Portuguese]
PAHO	Pan American Health Organization
PBC	performance-based contract
PLANASA	National Plan for Sanitation, in Brazil
PLANSAB	National Plan for Basic Sanitation, in Portuguese
PPP	public–private partnership
PRC	People's Republic of China
PUB	Public Utilities Administration of Singapore
RBF	results-based financing
RWI	Rural Welfare Institute
SARS-CoV-2	severe acute respiratory syndrome coronavirus 2
SDG	sustainable development goal
SFPUC	San Francisco Public Utilities Commission
STP	sewage treatment plant
SUDS	sustainable urban drainage system
UASB	up-flow anaerobic sludge blanket reactor
UN	United Nations
UNDP	United Nations Development Programme
UNEP	United Nations Environment Programme
UNESCO	United Nations Education, Science and Culture Organisation
UN-Habitat	United Nations Human Settlements Programme
UNICEF	United Nations International Children's Emergency Fund

UN-Water	'Coordination mechanism' in which the United Nations entities (members) and international organisations (partners) work on water and sanitation collaboratively
USD	United States dollar
WASH	water, sanitation and hygiene
WB	World Bank
WERF	Water Environment and Research Foundation
WHO	World Health Organization
WRF	Water Research Foundation
WWAP	World Water Assessment Programme
WWTP	wastewater treatment plant

doi: 10.2166/9781789064049_0001

Introduction

Can women bring a fresh perspective to the provision of water services? We believe so, and thus we have produced this book. This book is an attempt to compile with a critical and complementary point of view challenges, solutions and dilemmas we (men and women) face to achieve sanitation[1] for all. Our aim is to show that women not only have different views but also organise information differently, arriving at different conclusions. Furthermore, our aim is to show that women not only think differently, but also differ in thinking and decision making.

The provision and well-functioning of sanitation services is more challenging than for drinking water services. The state of progress reported in Chapter 1 confirms this idea: sanitation service coverage is lower than drinking water and progress to advance its procurement is slower. The 'sanitation ladder'[2] produced by the WHO–UNICEF joint monitoring programme, also described in Chapter 1, is an example of the complexity of the subject worldwide. The differences in the levels of the sanitation services of this ladder, as explained by experts and organisations, are due to the diverse political, economic, social and cultural contexts. Simply put, the ladder shows that inadequate sanitation is closely linked to poverty and inequity.

In high-income countries, both drinking water and sanitation services have progressed at the time when many other countries were still colonialised or were economically dependent. To fill the current gap, important investments

---

[1] Sanitation has many definitions, as will be discussed throughout the text. All are fine, but each reflects the social inequity that the differences in the services provided prevail, notably in the Global South. For this reason, we did not select a specific one.
[2] The JMP sanitation service ladder was developed to benchmark and compare service levels across countries (WHO-UNICEF JMP, 2016).

are needed and it is necessary to find ways to maintain facilities that function adequately. It is hard to convince politicians to invest in sanitation, as the subsector is frequently a source of complaints and is often perceived as an unpleasant subject to talk about publicly. The economic losses due to premature deaths, health care costs, productivity losses and time lost due to the practice of open defaecation are estimated to be of 2.5% of the mean global gross domestic product, a value that can attain 7.2% in some countries (World Bank, 2023a). These figures increase with epidemic outbreaks; losses in income from trade and tourism; the impact of unsafe excreta disposal on water resources quality and the long-term effects of poor sanitation on early childhood development (WWAP, 2015).

Society puts more pressure on the provision of drinking water as, after all, nobody can live without water. The way in which this service is provided, culturally and technologically, is relatively well standardised and politicians react quickly to its request as it soon reflects on votes. On the contrary, sanitation, particularly in rural areas and informal settlements of the Global South, consists of a series of services provided not only by the government but also by different types of big and small (even unipersonal) enterprises, community organisations and, even, non-governmental organisations, through what has been called the sanitation chain. Additionally, and in contrast once again with the drinking water service, sanitation is placed downstream of the users.

How can we advance on a subject that:

- Mainly concerns middle- and low-income regions which have a long list of other pressing needs?
- Is, by far, a service much more complex to provide than drinking water, technically, socially and financially?
- Is not only *per se* a human right, but is indispensable to achieve many of the other human rights. Just for the 2030 Agenda, sanitation is needed to reach the goals on poverty, health, education, gender, water, equity, cities and sustainable environment?
- Is a continuously growing problem, due to population increase and the need to raise the level of the services up to the standard being provided in high-income regions?
- The necessary financial and human resources are insufficient and, there is simply no political will to address the challenge?

Recognising the need to have new approaches for the provision of sanitation services, this book presents the current situation on sanitation, the challenges and proposed approaches to advance on this topic, especially in the Global South using the experience of women who have worked on the subject. It analyses the government structure and policy on sanitation, the role of policy and decision makers, better ways to realistically promote public participation and ideas for management and financing. The book is not a classic engineering book; it is for policy and decision makers who manage programmes, projects and systems. Thus, it does not cover the design, operation and maintenance of wastewater treatment systems. In addition, it contains several case study boxes, because this is the way in which women frequently discuss their ideas with others, by presenting and illustrating their ideas with concrete examples. Our effort is timely as according to the Glass report 2021/2022 (WHO, 2022a) only nearly a third of countries have elaborated their national plans and strategies on sanitation. During the production of this book, we noted that lots of solutions and points of view come from wealthier societies which try to address the needs for poorer ones, in some cases, without fully understanding the social and cultural context. Thus, even if we do not always provide alternatives, we, at least, intend to guide readers to reflect on the different challenges of sanitation in their local context.

Achieving 'Sanitation for All' implies special efforts of the few sanitation policy and decision makers working worldwide, but if they succeed it could change the life of many people in the world. Thus, it is an effort that cannot be performed alone; it is necessary to engage all of us: politicians, other sectors, donors and the society at large, including and foremost women.

doi: 10.2166/9781789064049_0005

Chapter 1

Global sanitation, situation and challenges: why do sanitation services advance more slowly than those for drinking water?

KEY MESSAGES:

- Sanitation is key for the well-being and daily life of people; therefore, it is a human right.
- Lack of sanitation negatively impacts health, education, gender equity, the environment and economic growth. In terms of health and education, a lack of sanitation compromises the fulfilment of the associated sustainable development goals (SDGs).
- Sanitation coverage and the quality of the associated services are an indicator of inequity, as are all of the WASH (water, sanitation and hygiene) subsectors.
- Sanitation coverage is low in middle- and low-income regions, and its progress is slow. It is unlikely the 2030 Agenda on sanitation targets will be met as this will require a five-fold increase in the current progress rate (16-fold in the Global South[1], 15-fold in fragile contexts; JMP, 2023).

---

[1] In this book, we are avoiding the use of the terms 'developed' and 'developing' countries, unless it is a quote. Instead, we are using, according to the context, the terms 'low-, middle- and high -income countries' and 'Global North and Global South'. Language shapes the way we perceive the world. The terms 'developing' and 'developed' countries construct a false narrative of the development concept that has been used to justify actions and policies grounded on the assumption that pure economic growth leads to a reduction in human poverty; however, 'development' has often led to the destruction of the natural environment and social relations. Additionally, it does not recognise the proper management Indigenous populations undertake of their environment. The first option used is to make it clear that differences arise because of the lack and accessibility to economic wealth. The Global North and Global South terms avoid the idea of 'developing countries' needing to become 'developed', but we must remain conscious of their binary nature. Moving away from juxtaposing terms could lead to language which is better placed to recognise who is considered an expert, whose knowledge is valued and who is looked at for solutions.

- Achieving SDG 6, notably target 6.2 'for all' implies, including refugees, asylum seekers, stateless persons and internally displaced persons. In these situations, the approaches for the implementation of sanitation services must also consider the weakness of the institutional context and the deficiencies and vulnerabilities suffered by everyone living within a region/country.
- The provision of sanitation services is very different from the provision of drinking water services. For sanitation services, the different social, cultural and environmental conditions are determinant to complete the chain of services to manage wastewater and faecal sludge and to reuse by-products. Also, for its provision an ample set of available technologies provide very different quality of services, notably from the comfort aspect. And most important of all, sanitation services are placed downstream of the users and their provision involves a varied set of stakeholders.
- The provision of sanitation is a complex subject with different evolving definitions used by many organisations. This leads to difficulties communicating with stakeholders and politicians, who are needed to support the entire process.

1.1 CURRENT PROGRESS ON SANITATION

According to the Joint Monitoring Programme (JMP, 2023), only 57% of the world's population has access to safely managed sanitation, that is, 3.5 billion people (Figure 1.1). However, 1.9 billion people lack even basic services, for 570 million the access to the service is limited, 545 million have an unimproved sanitation service and 419 million practise open defaecation (see Definitions at the end of the book).

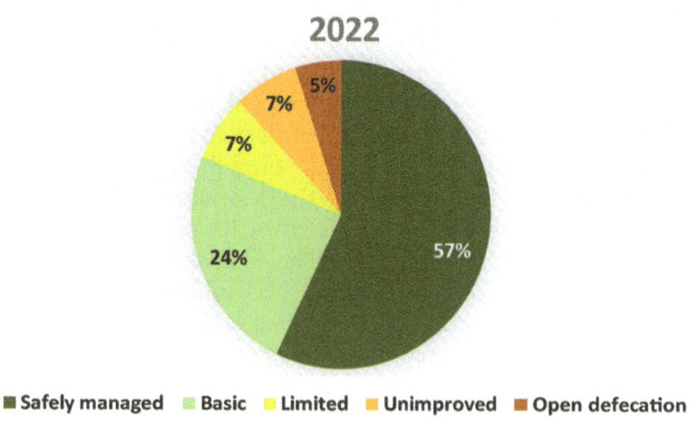

Figure 1.1 Access to different types of sanitation services and open defaecation practice worldwide in 2022 (*source:* JMP, 2023).

1.1.1 Non-sewered sanitation (NSS) and on-site (OSS) sanitation systems

Since 2000, the population with sewer connections has been increasing at an average of 0.41% persons annually. Growth in on-site systems has been faster for septic tanks, at 0.54% persons annually (JMP, 2023). In urban areas, due to population growth, the proportion of the population with sewer connections remained almost constant between 2000 (62%) and 2022 (63%), even though new facilities have been provided. For urban populations, the proportion of people using septic tanks increased from 15 to 22% (JMP, 2023). In fact, in 2022, globally, more people used on-site sanitation systems (OSS)[2] (46%) than sewer connections (42%), and most safely managed sanitation services were among households with sewer connections (33%), rather than on-site facilities (24%). There is limited updated information on the specific type of technology used. However, Figure 1.2 provides some indication.

Having access to basic sanitation is not the only challenge, as there are at least 2,700 million persons demanding proper management of faecal sludge, out of which around 450 million live in Africa. The global figure will increase to nearly 5,000 million people in 2030 (Cairns-Smith *et al.*, 2014; JMP, 2023; Peal *et al.*, 2014a, 2014b; Strande *et al.*, 2018).

A study of 12 cities across Africa, Asia and South America showed that, in many cases, faecal sludge remains buried in NSS–OSS, even if the content overflows polluting water courses or the soil (Mills *et al.*, 2014; Strande *et al.*, 2018). Due to a lack of funds to extract and safely dispose of the content, the service to empty toilets/latrines is delayed as long as possible

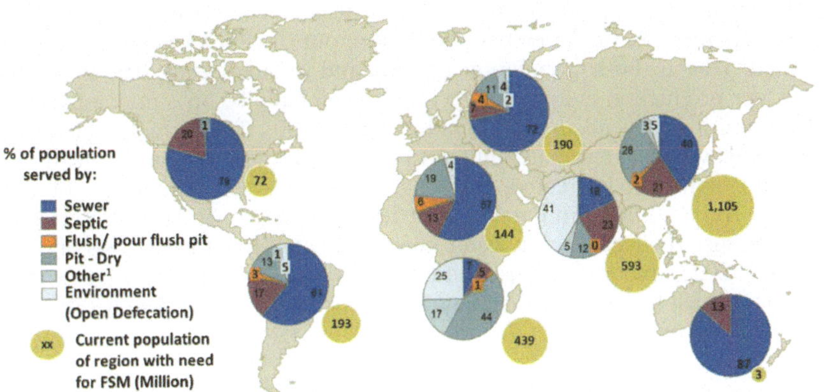

1. Open pits, pits without slabs and composting toilets included in "Other" as these do not need FSM (open pits/ pits without slabs covered up when full)

Source: UN JMP sanitation data, BCG analysis

Figure 1.2 Share of population served by different sanitation technologies, by region (*source*: Strande *et al.*, 2014).

---

[2] For OSS facilities to be counted as safely managed sanitation they need to ensure that at least excreta are well contained, and not discharged to surface water or soil, threatening human health and the environment (JMP, 2023).

(Jenkins *et al.*, 2015) and when performed, most of the sewage, faecal sludge and septage extracted is sent untreated into open drains, peri-urban fields or to poorly or non-functioning wastewater treatment facilities.

There are at least three main differences between sanitation and drinking water services. Firstly, for the provision of sanitation services, the different social, cultural and environmental conditions are determinant to complete the chain of services required from the users to the disposal into the environment and the reuse of by-products. Secondly, there is an ample set of technologies that are all applicable but provide a different degree of comfort for users and may or not be socially acceptable. Thirdly, drinking water services and facilities exist upstream from users and are basically managed by the government and eventually, at least partly by private companies; for sanitation these services are placed after the users and have to be built and operated between the government and a varied set of stakeholders. These characteristics partly explain the higher complexity in providing sanitation services, but also an ample set of constantly evolving definitions that are used by different organisations to describe the processes involved, as well as the difficulty in communicating the tasks and support needed from politicians and stakeholders.

1.1.2 Wastewater treatment

There is no global database on wastewater statistics, and the available information is not homogeneous. The available data are mostly from middle- and high-income countries and there is a clear underestimation of the amount of wastewater produced (UN-Habitat & WHO, 2021). This information reports a total (industrial and domestic) production of wastewater of nearly 132 million m^3 annually for 22% of the global population, out of which only around 32% is treated. Due to both population growth and an increasing use of water, wastewater production is expected to increase globally by 56% in comparison to the amount produced in 2015 (Qadir *et al.*, 2020). The coverage of sewer connections, which is basically limited to cities, does not necessarily mean the wastewater collected is treated. In fact, treated collected wastewater ranges from <1% to over 99% in different countries (JMP, 2023). To be considered as safely managed, collected wastewater must be treated to, at least, a secondary level (JMP, 2023), independently of whether it is disposed into the soil or water bodies, for which the treatment and management has to be completely different (Jiménez Cisneros, 1995).

1.2 SANITATION DISPARITIES

1.2.1 Income and regions

There are pronounced disparities on sanitation coverage per capita income, region (Figure 1.3) and country (Figure 1.4). Differences are not only observed in the coverage of safely managed sanitation systems, use of basic sanitation facilities, the emptying and safe disposal of OSS content, the collection and treatment of wastewater but also in the efficiency and quality of all these services which is associated with the availability of sufficient operational funds.

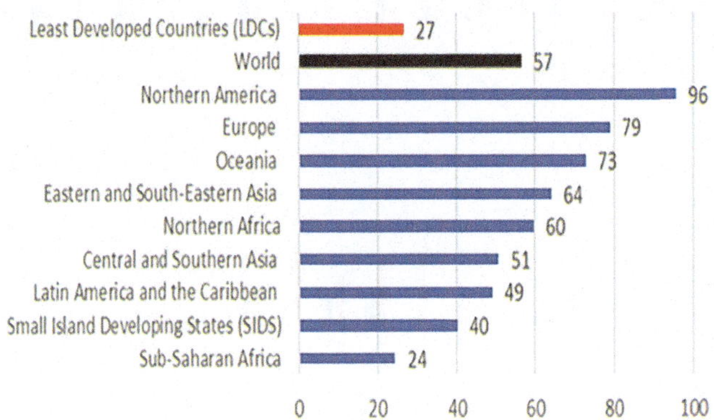

Figure 1.3 Share of the population using safely managed sanitation (*source*: with information from JMP, 2022).

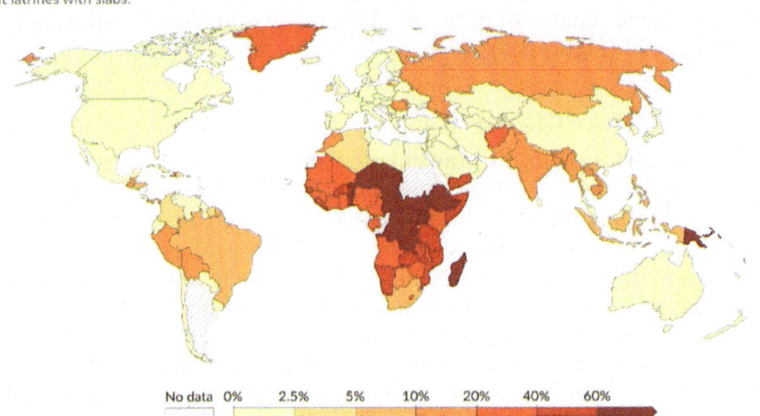

Figure 1.4 Share of the population without access to improved sanitation per country in 2022 (*source*: our world data, 2023, which uses information from JMP, 2022).

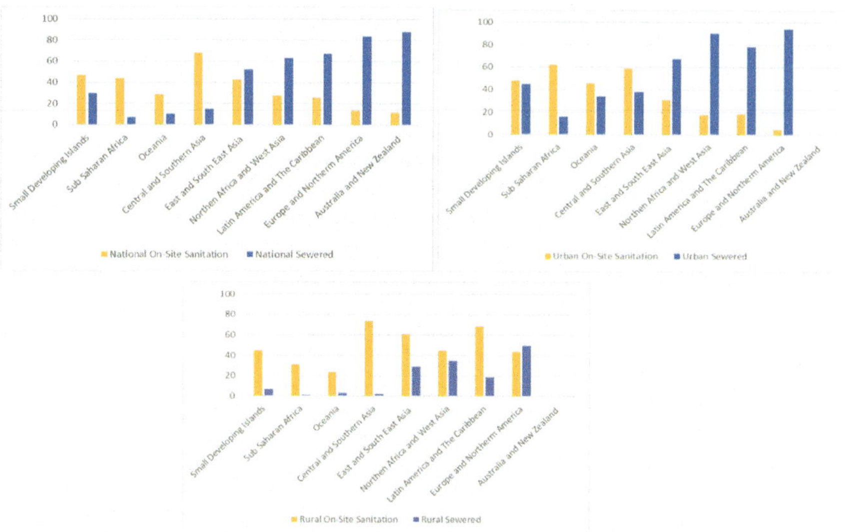

Figure 1.5 Variation of the distribution of the population using OSS and NSS for rural and urban areas and at national level in 2020 (%) (*source*: WHO, 2020).

1.2.2 Urban and rural areas

Around two-thirds of people who lack basic services live in rural areas. Nearly half of them reside in sub-Saharan Africa. The use of sewers is a notorious difference between urban and rural settings; these differences are more pronounced in some regions than others. This is illustrated using information available from 2020 (Figure 1.5).

1.3 UNSAFE SANITATION IMPACTS

Poor sanitation is linked to the transmission of diarrhoeal diseases such as cholera and dysentery, as well as typhoid, intestinal worm infections and polio. It exacerbates stunting and contributes to the spread of antimicrobial resistance. Open defaecation (practiced by 6% of the world population) is the main contributor (De Shay *et al.*, 2020). In low-income countries, 5% of all deaths are associated with unsafe sanitation (Ritchie *et al.*, 2019), and it is a problem in specific countries mostly from sub-Saharan Africa and southern Asia (Figure 1.6). Poor sanitation affects individual's mental well-being and personal safety, especially for women and children, who risk bodily exposure, harassment and violence when practicing open defaecation (De Shay *et al.*, 2020).

Water pollution has worsened in almost all rivers in Africa, Asia and Latin America due to a low sanitation coverage. The deterioration of water quality is expected to further escalate over the next few decades, compromising human and environmental health, as well as sustainable development in several countries. Faecal coliform loadings to rivers are high or very high in many

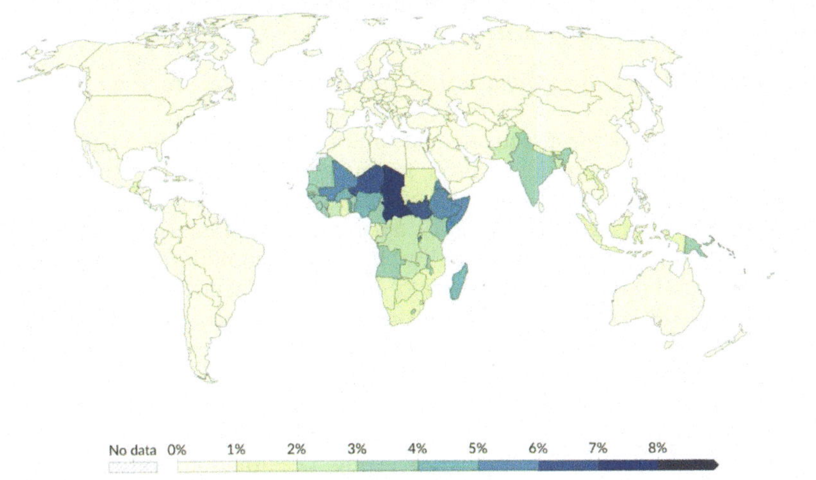

Figure 1.6 From Ritchie *et al.* (2019) who used information from IHME, Global Burden of Disease (2019).

regions because of the lack of wastewater treatment. However, the solution is not only to build sewers or OSS facilities, but also to treat the collected wastewater and to safely manage faecal sludge. Globally, the most prevalent water quality challenge is nutrient loading, which, depending on the region, is often associated with the loading of pathogens, biodegradable organic matter and chemical toxic compounds too. Sewage is one of the biggest contributors to ocean pollution, with more than 80% of global sewage flowing untreated into our seas (Big Blue Ocean Clean up, 2023). Wastewater diverted into the ocean is a source of plastics too.

Annual economic losses due to poor sanitation are equivalent to between 1 and 2.5% of gross domestic product due to premature deaths, health care costs, productivity losses and the time lost due to the practice of open defaecation. The actual cost could be much higher when considering the costs of epidemic outbreaks; income losses from trade and tourism; impact of unsafe excreta disposal on water resource quality and the long-term effects of poor sanitation on early childhood development (UN-Water, 2015a).

1.4 SANITATION AND THE 2030 AGENDA

Access to safe sanitation is essential for reducing diseases and deaths from infectious diseases, preventing malnutrition and ensuring dignity. It also improves cognitive development and increases working days and economic development (Freeman *et al.*, 2017; Sclar *et al.*, 2017; Speich *et al.*, 2016; Wolf *et al.*, 2014). This is why sanitation is part of sustainable development goal (SDG) 6 'Ensure availability and sustainable management of water and sanitation for all' (UNDESA, 2015). Among the eight targets of SDG 6, those closely related to sanitation are:

- **Target 6.2** *End open defaecation and provide access to sanitation and hygiene*
  By 2030, achieve access to adequate and equitable sanitation and hygiene for all and end open defaecation, paying special attention to the needs of women and girls and those in vulnerable situations.
- **Target 6.3** *Improve water quality, wastewater treatment and safe reuse*
  By 2030, improve water quality by reducing pollution, eliminating dumping and minimising release of hazardous chemicals and materials, halving the proportion of untreated wastewater and substantially increasing recycling and safe reuse globally.
- **Target 6.7** *Expand water and sanitation support to developing countries*
  By 2030, expand international cooperation and capacity-building support to developing countries in water- and sanitation-related activities and programmes, including water harvesting, desalination, water efficiency, wastewater treatment, recycling and reuse.
- **Target 6.8** *Support local engagement in water and sanitation management*
  Support and strengthen the participation of local communities in improving water and sanitation management.

Additionally, for target 6.2, which is specifically for sanitation, there are other two targets referring to the means of implementation of the entire SDG 6 apply to sanitation, these are:

- **Target 6.a** By 2030, expand international cooperation and capacity-building support to developing countries in water- and sanitation-related activities and programmes, including water harvesting, desalination, water efficiency, wastewater treatment, recycling and reuse technologies.
- **Target 6.b** Support and strengthen the participation of local communities in improving water and sanitation management.

Achieving SDG target 6.2 'for all' implies, including refugees, asylum seekers, stateless persons and internally displaced persons. For which, the approaches to implement the sanitation services must consider the weakness of the institutional context and the characteristics, deficiencies and vulnerabilities of the different population sectors (UN-ESCAP, UN-HABITAT & AIT, 2015; UNICEF & WHO, 2020).

Sanitation is particularly important to fulfil SDG 4 to 'Guarantee inclusive, equitable and quality education and promote lifelong learning opportunities for all'. Target 4a focuses on school infrastructure and the need to 'build and adapt educational facilities that are sensitive to the needs of children and people with disabilities and gender differences, and that provide safe, non-violent learning environments, inclusive and effective for all', for which one of the indicators

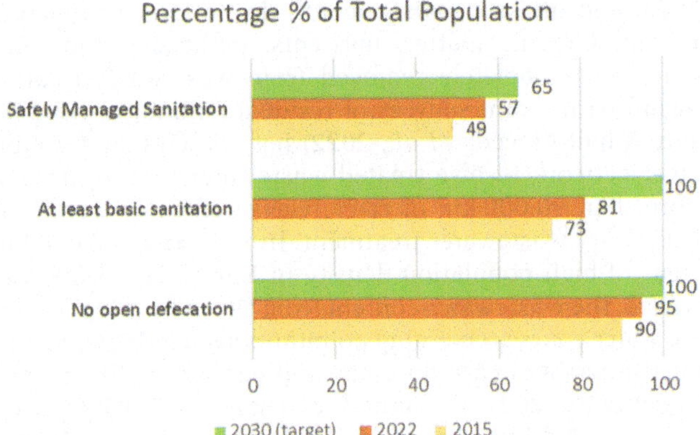

Figure 1.7 Progress achieving sanitation for all in 2022 compared to 2015 and target for SDG 6.2 for 2030 (*source*: with information from JMP, 2023).

used to measure progress towards this goal is the proportion of schools with access to basic sanitation facilities separated by sex.

Despite the increased political attention and financial support on sanitation, there is still a long way to go to achieve the corresponding targets by 2030 (Figure 1.7) as this will require a five-fold increase in safely managed sanitation on current progress rates (16-fold in least developed countries, 15-fold in fragile contexts; JMP, 2023). This seems challenging, but it looks much more complicated to achieve when considering there is a need to end open defaecation and transition unimproved, limited and basic sanitation into safely managed and sustainable sanitation practices.

1.5 THE HUMAN RIGHT TO SANITATION

Five years before the adoption of the 2030 Agenda for Sustainable Development (UNDSSA, 2015), the United Nations General Assembly's resolution 64/292 (UNGA, 2010) and the Human Rights Council's resolution 15/9 (UNHRC, 2010) recognised the human rights to water and sanitation (HRtWS). However, Brown and Heller (2017) pointed out that although sanitation was recognised as a human right by the United Nations member states, due to its complexity to be achieved, sanitation 'is a concept still under construction that needs to be approached and interpreted in a consensual way by all involved'. Beyond, recognition of the HRtWS has enabled closer dialogues between governments, civil society groups, service providers and development practitioners on how to integrate human right principles into policies and plans, incorporating the SDG language 'Leave no one behind' (Heller *et al.*, 2020).

1.6 SANITATION CHAIN APPROACH AND CIRCULAR ECONOMY

An issue that still requires reflection is that even the more sophisticated ways by which we currently have to provide sanitation (collection of waste water in sewers and its 'safe management') do not ensure full protection of the environment. Organic matter, nutrients, pathogens and other specific compounds are not completely removed from wastewater treatment plants (WWTPs), resulting in point sources of residual pollutants in surface waters. A recent study (Ehalt Macedo *et al.*, 2022) has shown that 1.2 million km of the global river network receive treated wastewater from upstream WWTPs. Of these, more than 90,000 km of river receive effluents from WWTPs that only provide primary wastewater treatment. In more than 72,000 km of rivers, mainly in areas of high population density in Europe, the USA, China, India and South Africa, the wastewater content from WWTPs is higher than 10%. In many of these water courses emerging pollutants such as human and veterinary pharmaceuticals, personal care products and endocrine disrupters have been found (Branchet *et al.*, 2021; Garduño-Jiménez *et al.*, 2023; Gaw *et al.*, 2014; Khan *et al.*, 2020; Madikizela *et al.*, 2017; Mezzelani *et al.*, 2018; Mo *et al.*, 2022; Peña-Guzmán *et al.*, 2019; Świacka *et al.*, 2022). An option to address this new challenge is to move to a circular economy concept, reducing the amount of wastewater produced, improve its quality by avoiding use and discharge of residual pollutants and reusing water in an intentional and planned way; after all, water is a resource that is neither destroyed nor created and we have always been reusing it.

doi: 10.2166/9781789064049_0015

Chapter 2

Sanitation: an unavoidable public responsibility

KEY MESSAGES:

- The 'government for sanitation' has four components: the first providing the ideology (the policy); the second providing operational structure (institutional framework); the third (closely related to the second) providing the rules for operation and the way in which the other three are linked and, the fourth, which is the social component, that is the stakeholders and partners.
- There are several definitions of sanitation, which reflect the complexity of this service and its close link to poverty, inequity and cultural context.
- Creating an enabling environment to develop, improve and implement a sustainable sanitation for all requires the design, operation and maintenance, in a step-by-step approach, of an operational government structure able to deal with the entire sanitation service chain.
- For the provision of sanitation several sectors and governmental entities (institutional framework) at the national, regional and local levels intervene. It is important for all – entities and sectors –, to have clear operational mandates and coordination mechanisms, defined both through the legal and institutional frameworks.
- Policy and decision makers do not play the same role as academics and researchers; each one has their own niche. The first needs to act with the available knowledge and information, considering political aspects and under a specific time frame; the second provides knowledge and information and suggests decisions for which they will neither be responsible nor be accountable for.
- To efficiently implement sanitation, it is imperative to define who will have the coordination role. This must be done keeping in mind that

implementation takes place at the local level and the local government has direct responsibility. Nevertheless, the local government frequently lacks funds, human resources and political support.

- Regulatory mandates and functions are often more clearly defined for water supply than for sanitation. Contrary to water supply services, sanitation services are placed downstream of the users and not all services are provided by the government. Indeed, several entities participate in the sanitation chain, including the water sector government, an ample set of institutions, private enterprises, social and international organisations.
- Challenges to implement sustainable sanitation for all include technical, socio-cultural, political, financial and readiness to deal with future changes throughout the sanitation chain.
- Innovative approaches have been developed to provide sanitation strategies which policy and decision makers could consider in their plans.
- The management of wastewater as a resource in the context of circular economy and using natural-based, environmentally friendly and hybrid solutions together with the planning of smart cities are new approaches that may lead to a better progress for sanitation.

2.1 SANITATION AND THE SANITATION CHAIN

There are several definitions for 'sanitation' depending on the country, organisation and approach (Wikipedia, 2024; https://en.wikipedia.org/wiki/Sanitation). It is beyond the scope of this book to analyse all of them. For the purpose of this book, the following working definition will be used: 'sanitation refers to the safe management of human excreta and wastewater and their disposal and includes the reuse of water and the recycling of byproducts'. The safe management comprises the collection, transportation and treatment before disposal or even revalorisation of wastewater, sludge, faecal matter and the material extracted from on-site sanitation (OSS) systems. The main objective for sanitation is to protect human health by avoiding the transmission of diseases especially through the faecal–oral route[1]. Sanitation also serves to protect the environment, notably the pollution of water bodies for which it is necessary to treat wastewater and material containing faecal matter before disposal or reuse.

A sanitation system or 'sanitation chain'[2] includes the facilities for defaecation, the collection, storage, transport, treatment and disposal or reuse of human excreta and wastewater, and reclamation of the associated by-products. Providing sanitation requires attention to the entire sanitation

---

[1] The faecal–oral route is a pathway of transmission of diseases through pathogens. The main causes of faecal oral disease transmission are the lack of adequate sanitation, poor hygiene practices and the non-hygienic management of food. Examples of diseases transmitted through lack of sanitation are cholera, helminthiasis (diseases caused by worms such as ascariasis), hepatitis and typhoid.

[2] In the literature, often the revalorisation of water or byproducts is termed 'sanitation value chain'. However, we do not differentiate between the 'sanitation chain' and the 'sanitation value chain', as nowadays the sanitation chain should simply consider reuse and reclamation practices.

chain, focusing not only on technical aspects but also on social, environmental and economic ones.

WHO-UNICEF JMP (2016) relatively recently recognised that there are several 'levels' of sanitation services across countries and developed a ladder[3] for sanitation listing practices well known in the Global South. Although the terminology and classification developed in this framework is complex and not yet universal, it is useful to show the different solutions to provide sanitation worldwide. The ladder is based firstly on classifying systems in terms of improved or unimproved. Improved sanitation facilities[4] are those designed to hygienically separate excreta from human contact. The next level is to consider whether the practice is safely managed; facilities should not be shared between households, and the excreta produced should either be: (a) treated and disposed of in situ; (b) stored temporarily and then emptied and treated off-site or (c) transported through a sewer with wastewater and then treated off-site. When the excreta from improved sanitation facilities are not safely managed then the practice is considered as being a basic sanitation service. The sanitation ladder, from the lowest to the highest level is as follows:

(1) Open defaecation: disposal of human faeces in fields, forests, bushes, open bodies of water, beaches and other open spaces or with solid waste.
(2) Unimproved: use of pit latrines without a slab or platform, hanging latrines or bucket latrines.
(3) Limited: use of improved facilities shared between two or more households.
(4) Basic: use of improved facilities which are not shared with other households.
(5) Safely managed: use of improved facilities that are not shared with other households and where excreta is safely disposed of in situ or removed and treated off-site.

2.2 SANITATION WITHIN THE WATER SECTOR

It is important to identify the proper placement of sanitation within the water sector, given it includes sanitation services *per se* which are part of the water services[5] and the reuse of water which is part of the administration of water, as reuse is a source of water. This is an issue, as in some countries the management of water resources and that of water services (drinking water and sanitation) are placed in different institutions (Box 2.1), affecting the way in which the sanitation chain is managed.

2.3 THE GOVERNMENT FOR SANITATION

An adequate government for sanitation (GWP, 2008; Lautze *et al.*, 2011; Özerol *et al.*, 2018; Rogers & Hall, 2003) has four components that can be arranged

---

[3] A classification that was made to be able to continue using the data that were collected in the past by the JMP (Joint Monitoring Programme).

[4] Improved sanitation facilities include flush/pour flush toilets connected to piped sewer systems, septic tanks or pit latrines; pit latrines with slabs (including ventilated pit latrines) and composting toilets.

[5] i.e. the services for the provision of drinking water and sanitation.

Box 2.1 Consolidating water resources and water services in a single institution: exploring advantages and disadvantages

Some governments have consolidated different ministries responsible for various water-related functions under a single institution. This is to promote a more coordinated and integrated approach for water management. By bringing together different functions of water such as for sanitation, drinking water, irrigation and environmental pollution, it may be possible to identify synergies and trade-offs between all the functions. In the context of sanitation, the management of water reuse and pollution control by a single institution leads to a more effective, comprehensive and sustainable management of water.

In 1989, Mexico consolidated the administration of water resources and services in a highly specialised institution, the National Commission for Water (Comisión Nacional del Agua, CONAGUA). CONAGUA includes the water use administration (including environmental 'use'), discharge permits for municipalities, industries and farmers, federal coordination of water services (drinking water, sanitation and reuse), management of federal irrigation and hydraulic infrastructure (channels, dams, water transfers, etc.), flood and drought risk management and the meteorological service. Its objective is to administer, regulate, control and protect national water with the participation of civil society to achieve the sustainable management of the resource. CONAGUA is a deconcentrated institution which promotes public participation through basin organisations where all stakeholders jointly prepare the local, regional and national hydrological plans (National Water Policy). In terms of water services, CONAGUA sets specific programmes and allocates funds to ensure the human rights to water and sanitation (HRtWS). It also oversees fund allocation to improve quality of the water services for all. The Ministries of the Environment, Finance, Social Welfare, Economy, Health, Agriculture, Livestock and Fishing, and the Forestry Commission and other relevant social and academic representatives are part of its technical advisory board (CONAGUA, 2023).

CONAGUA is sectorised within the Environment Ministry, but the economic and political power differences between the Environment Ministry and CONAGUA hinder the harmonisation between both organisations. The Ministry represents only one of the water's users (the environment) and performs only bureaucratic tasks, whereas CONAGUA oversees the protection of all the users of water, undertakes presidential tasks and builds important hydraulic federal infrastructure.

Malaysia consolidated its water and environment ministries in 2009 creating the Ministry of Energy, Green Technology and Water. This led to greater coordination between the water and environment sectors, resulting in more effective management of water resources and improving the environmental protection. Similarly, in the Philippines, the creation of the Department of Environment and Natural Resources led to more effective integration of environmental and natural resource management (Department of Environment and Natural Resources, 2023).

Consolidation of water management functions can lead to greater transparency and accountability, as there is a single ministry or department responsible for the entire water and sanitation sector. This could help to reduce duplication and fragmentation of functions, improving efficiency and effectiveness. Additionally, having a single point of contact for the water and sanitation sector helps to streamline communication and coordination between different stakeholders and partners (AFBD, UNEP & GRID-Arendal, 2020). However, there are potential drawbacks when reuniting water management functions. One concern is that combining functions could lead to a loss of expertise and focus on specific areas of water management, potentially leading to less effective policies and programmes. For example, the consolidation of the Department of Water Affairs and Forestry with the Department of Environmental Affairs and Tourism in South Africa in 2009 has been criticised for reducing the focus on water management and sanitation. Similarly, in India, the merger of the Ministry of Water Resources, River Development and Ganga Rejuvenation with the Ministry of Drinking Water and Sanitation has faced challenges due to conflicting priorities and a lack of clarity around responsibilities.

Another potential challenge is bureaucracy and red tape involved in decision making and implementation. This is because these processes may have to pass through more layers of approval and oversight. Furthermore, there may be conflicts of interest and priorities between different departments, potentially leading to less effective policies and programmes. For instance, in the USA, the Environmental Protection Agency (EPA) is responsible for setting water quality standards, but the Department of Agriculture (USDA) is responsible for providing financial support to farmers, including those whose activities may contribute to water pollution (USDA, 2024).

Careful planning, communication and ongoing evaluation are essential for ensuring that consolidation of water management functions leads to more effective and sustainable water management.

Source: partly, with information from AFBD, UNEP and GRID-Arendal (2020).

in different ways (Figure 2.1): a national sanitation policy, an institutional framework, a legal framework and the stakeholders. The national sanitation policy has the vision for the subsector and defines the goals for the service. It contains what must be done, the general procedures on how this should be achieved and who is responsible for different tasks. The institutional framework is the set of formal public organisations (in the water sector or other sectors, and private, public or social groups) functioning according to their

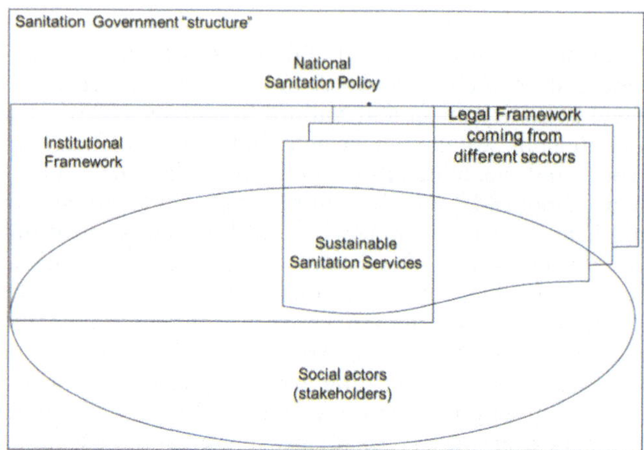

Figure 2.1 Components of the sanitation government.

own mandates. The legal framework includes laws, regulations, norms and standards specifically developed within the water sector and also by any other sectors. It has the rules for operation for all components and states the rights and responsibilities stakeholders have. The stakeholders comprise an ample set of varied public, social and private organisations and persons (individual or legal) that are participating along the sanitation chain as users, service providers or by being indirectly impacted by the service. In almost all countries these four elements already exist and function with different degrees of efficiency.

2.3.1 National sanitation policy

There are four core aspects to consider for the national policy on sanitation: (a) what should be the government's role; (b) what could be the role of third parties; (c) what model will be implemented and (d) the role for stakeholders and partners.

2.3.1.1 The government's role

Sanitation is a government responsibility, because the human and environmental health and the society welfare are an obligation that must be fulfilled independently of whether profits are generated or not. Some of the tasks the government is responsible for are:

- Ensuring the provision of sanitation to the entire population.
- Implementation and overall management of the sanitation chain; which includes the users receiving the service, those that treat wastewater, manage faecal sludge and reclaim water and by-products or reuse them, among others.
- Production of regulations and standards to provide the service.
- Supervision of all stakeholders and partners involved in the sanitation chain to assess the quality of the services and ensure the protection of human and environmental health.

- Keeping the public informed and promoting public participation throughout the entire sanitation service chain.
- Ensuring proper financing mechanisms are available and accessible to all and that the cost for sanitation is accessible to the entire population.

Some countries have decided to create independent and specialised institutions (sometimes public sometimes private) to perform the functions of regulation, supervision and implementation. The justification for this is to provide counterbalances. However, as discussed in Box 2.1, this does not lead to an efficient use of human and economic resources. Furthermore, when autonomous institutions are created for supervision (public, private companies or non-governmental organisations (NGOs)), frequently there is an uneven distribution of power and even of salaries. This leads to the supervising agency becoming a powerful organisation with professionals working at the desk level, far from daily problems, overseeing those having direct responsibility for providing the service (Cunha Marques, 2010). Using independent bodies for service implementation has increased costs; services are not always improved, new technology is not always developed, nor a full universalisation of the service is achieved. In general, the creation of private independent bodies may weaken the expertise of the public bodies as they usually provide better salaries, leading experienced people to move from public entities to the independent ones (Neves-Silva *et al.*, 2023; Post & Athreye, 2016; World Bank, 2023b; Wu *et al.*, 2016). Therefore, when dividing functions and creating agencies, it is important to consider the total operating costs, accountability, efficiency, the size of services to provide and the country/regional characteristics.

Although core functions for sanitation cannot be delegated, some specific tasks may be performed successfully by third parties. Some examples of outsourcing tasks are the evaluation of the role and amount for sanitation tariffs, bench marking, conflict resolution and performance of financial mechanisms. Nevertheless, to achieve good results, these schemes face the challenge of perfectly defining the roles and responsibilities of each party (Halpern & Trémolet, 2006; IWRM Action Hub, 2023).

2.3.1.2 *The role of third parties*

The sanitation services usually are provided by 'water utilities'[6]. Policy and decision makers can decide, if the country's general administrative legal framework allows privatisation, if they would like support from private entities for the provision of the service and up to what extent (Table 2.1). This is known as private participation, as the government remains fully responsible and accountable for the provision of the service. Private participation may go from planning or designing infrastructure, to its construction and operation. Private entities include private companies, NGOs, international or national private or social organisations and professional associations. None of them are accountable to the public (GWP, 2008), although they are accountable to their boards or the organisation they belong to, but none have the responsibility to work for the public good.

---

[6] In some countries of the Global South (or Majority World) in rural settings the services are provided by community organisations that are independent of the government.

Table 2.1 Advantages and disadvantages of public and private participation.

Model	Advantages	Disadvantages
Full private participation (designs, builds and operate)	Faster response to the increase and improvement of sanitation services. More flexibility to implement innovative solutions.	More expensive than public provision. The increase in the coverage is fast provided funds are available. Requires a good definition of the tasks to be performed to produce a contract, followed by close supervision. It has been frequently rejected by the public, mostly in the Global South.
Full public	Usually, slow reply to user's requests.	Lower costs. Higher transparency as the government is fully accountable and there are fewer layers to investigate.
Public management with private participation for specific tasks	If well designed and operating well it may combine the advantages of both models.	More complex to operate.

Source: partly, with information from GWP (2008), WB (2023b), OECD (2009) and Gupta and Pahl-Wostl (2013).

Private companies are driven by profits; for the water sector, this means they are not motivated by addressing new public demands, introducing new technologies, reducing greenhouse gas emissions, further protecting the environment or adapting the infrastructure to climate change impacts. Furthermore, they do not look to invest their profits to finance extension on coverage of the service nor to address inequities. Nowadays, it is recognised that: (a) the water 'industry' is a capital-intensive industry with a low and slow returns for private investments, and (b) the water infrastructure required cannot be subject to a policy of recoverable costs as wastewater tariffs rarely recover the costs of the service (OECD, 2015, 2019). This means that the government must continue investing to maintain, increase, improve and resolve inequities on the provision of the services.

2.3.1.3 Non-governmental sanitation stakeholders and coordination with public institutions

As mentioned, the sanitation service chain is placed after the users[7] and it necessarily involves public, social and private participation. Thus, unlike the water supply service, for sanitation the non-public stakeholders are not only relevant but also needed. These non-public stakeholders come from the private

---

[7] For drinking water services, all the government has to do to address users' needs is done before the product (water) reaches the user. In contrast for the sanitation chain, all the services that have to be provided are placed once the users dispose of used water or faecal matter; the series of services, notably when reuse is involved, are placed downstream of the users and not all are the responsibility of the water government sector.

sector (from big enterprises to one -person business), the civil society, national or international organisations from the water or other sectors, community-based organisations, NGOs, universities, research centres and so on (GWP, 2008). All of them contribute independently and as part of institutional committees/commissions created to support the government's role.

This wide variety of actors require strong and efficient coordination mechanisms, in which the sanitation authorities must play a key role, for which it is important to understand all, government and non-government parties, their role, responsibilities and rights, as well as the role the coordination agency plays (Table 2.2). One non-trivial aspect is that among all stakeholders and partners only those from the government are accountable, that is why the leading and coordination role is governmental. Empowering any other actor, for example, an enterprise, social association, academia or NGOs, is a way of privatising the service, as there is no longer an entity responsible to veil for the public good.

Table 2.2 Essential functions to be fulfilled by organisations and actors (stakeholders) to ensure sustainable sanitation and water management.

Organisation and Actors	Functions
National authorities	National coordination role. Set the general conditions to achieve and implement the universal right to water. Define if private participation is allowed and under which terms and for which activities. Supervise operation of the sanitation services. Report the national figures on sanitation. Analyse the national information and set benchmarks for the services, with support from other stakeholders. Set international financing lines for sanitation, when appropriate.
Regional authorities	Regional coordination role. Provide financial and political support to local authorities to fulfil their sanitation tasks. Define the regional legal framework to provide sanitation. Provide administrative procedures and practical details on how to implement sanitation tasks. Participate in the process to get international funding.
Local authorities	Local coordination role, notably for on-site implementation. Identify sanitation priorities and help defining regional and national goals. Responsible for the provision of sanitation and thus directly accountable to citizens. Directly supervise companies and/or public institutions providing sanitation services. Support and provide guidance to set the regulation of the services. May raise funds.

(Continued)

Table 2.2 Essential functions to be fulfilled by organisations and actors (stakeholders) to ensure sustainable sanitation and water management *(Continued)*.

Organisation and Actors	Functions
Regulatory and enforcement bodies	Establish the roles and define tools for the sustainable management of sanitation.
Service providers (include government departments, municipal councils, public corporations, private sector companies, community-based organisations, farmers group and others	Provide components of the entire sanitation service chain, according to the legal framework.
Private sector	Provide the service as part of the service providers. Participate in the financing, provision of equipment or construction of facilities in line with the legal framework.
Civil society institutions, NGOs and community-based organisations	Provide the service as part of the service providers. Play an advocacy role in the formulation of policies. Raise awareness, undertake communication campaigns and mobilise local communities. Develop new models and tools for sanitation or support their testing in the field. Advocates for other specific purposes directly or indirectly related to sanitation such as environmental protection, and other private interests.
Universities and research centres (comprising public and private education and research centres)	Key in providing information at the local level, specific to countries and communities, on the impact and progress of sanitation programmes. Contribute to the testing of new models and technologies on-site. Assist in assessing the progress made by sanitation programmes, providing measured evidence. Provide explanations as to why problems arise and propose solutions. Assist in the selection and proper operation of different technologies. Assist in the integration of local and traditional knowledge to conventional practices. Assess the impact of sanitation programmes on health, education, gender equity, food security and so on.

Source: partly, with information from Peters (2011), Gupta and Pahl-Wostl (2013) and WHO (2006).

2.3.1.4 Models for the provision of sanitation

There are three models to provide sanitation: fully centralised, fully decentralised and a mixture of the two.

> *Fully centralised model.* This model also applies to the supply of water for which extended water networks are designed to transport water over great distances, often requiring a significant amount of energy to pump water to

urban centres. A piping network to discharge the used water ('wastewater') away from urban centres usually mirrors the water supply piping network. This 'traditional'[8] way of providing sanitation employs centralised systems with extensive piping networks and big wastewater treatment plants (at least when available, which occurs in middle high- and high-income regions). This linear historical approach of 'water in and water out' became the norm during the late 19th century and continued throughout the 20th century. It is practiced for many good reasons: one of which is that it facilitates having a highly specialised agency. Today, many of the centralised water and sanitation systems around the world pose significant and increasing economic, social and environmental costs to the communities they serve. Large sewers and wastewater treatment plants have become a source of punctual pollution entailing impacts of diverse intensity and nature.

Fully decentralised model. It is also known as a localised solution model. Frequently, this model jointly manages water and wastewater, mostly in small and/or remote communities. In the last decade, decentralised systems have increasingly been successfully applied in medium- and low-income urban areas and in rural areas (Box 2.2).

The idea behind the fully decentralised model for sanitation is to build and manage systems that are more resilient and sustainable. It comprises the capturing and treating of wastewater/faecal sludge on-site. It can also include the reuse of water for diverse activities such as green area irrigation, car washing, replenishment of fountains and so on, or contributing to food security by reusing water to grow crops. The faecal sludge composted can also be used to improve soil productivity. In principle, OSS systems also have the potential to: (a) reduce energy needs as less pumping is needed when reuse is performed on-site, (b) respond to rapidly growing communities as compared to public investments in centralised infrastructure, (c) provide additional benefits through resource recovery opportunities, such as producing thermal energy, (d) mobilise and engage the community, civil society and corporations in the management of water and (e) reduce capital expenditures for utilities and mobilising private investment for public benefit (Massoud *et al.*, 2009).

Despite these advantages, the adoption of a fully decentralised model for sanitation still presents challenges and its implementation demands processes that need to be further developed. Box 2.3 describes such processes for the Japanese case presented in Box 2.2.

The centralised–decentralised model. Cities are mostly built using conventional sewerage models, so water and sanitation services tend nowadays to be expanded using a combination of both centralised and decentralised models. Also, because of the increasing financial liability of older centralised systems and the lack of land to bury large wastewater collection pipe networks in currently underserved urban centres, there is significant interest in decentralised or hybrid systems (World Bank, 2017). Table 2.3 provides a comparison among the different models for sanitation.

---

[8] Traditional in the context of developed countries.

Box 2.2 Decentralised sanitation in rural areas: lessons from Japan's success story

Japan has achieved remarkable success in implementing on-site wastewater systems, particularly in rural areas. This success can be attributed to meticulous planning, stringent regulations and a committed focus on human resource development and environmental management. The implementation of the programme is based on five pivotal aspects: infrastructure development, regulation and compliance, operation and maintenance (O&M), human resource development and environmental monitoring and management.

(1) *Infrastructure development*: In Japan, the establishment of over 1,000 night soil treatment facilities prior to the development of sewerage treatment plants served as a critical aspect of this foundation. These facilities handled sludge treatment from the Johkasou on-site systems. Simultaneously, large-scale on-site systems in commercial buildings were carefully managed to ensure they are adhered to the necessary standards and did not lead to environmental issues.

(2) *Regulation and compliance*: Japan has strict regulations guiding the O&M of on-site systems. The On-site System Act mandates frequency of O&M, requiring technical supervisors when user levels attain a certain threshold. Large-scale on-site systems, especially those situated in office buildings and commercial spaces, are also rigorously monitored under multiple laws to ensure environmental compliance.

(3) *Operation and maintenance*: Maintenance checks, adjustments and disinfectant replenishments occur regularly in Japan, contributing to the sustainability of advanced on-site systems. The process also includes environmental accountability wherein the effluent water quality of all on-site systems is evaluated periodically to confirm their compliance to environmental standards.

(4) *Human resource development*: By implementing certification and training systems for on-site system operators, Japan ensures the maintenance of on-site systems by qualified professionals. Additionally, the nation emphasises awareness programmes and resource development in this sector. Institutions such as the Japan Education Centre for Environmental Sanitation play a significant role by offering exams and training courses, preparing over 3,000 technicians each year.

(5) *Environmental monitoring and management*: As part of their regulatory compliance, all on-site systems are subjected to annual inspections by a designated agency. The inspections cover not only the effluent water quality but also frequencies of desludging and maintenance works. This careful monitoring allows for the continued accountability of system operators and users.

Source: ADB (2021a).

Box 2.3 Success of the Japan's on-site system implementation: a public matter

A substantial portion of the Japanese population depends on OSS systems, particularly in medium and small cities. However, the problem with these systems is that they remain a 'no person's land': 'politicians and government officials tend to think that they are not a public matter, while, on the other hand, most people do not care about them'. So, making on-site systems a public matter is fundamental when considering regulatory frameworks for citywide inclusive sanitation (CWIS). This aspect is as important as the technical and regulatory solutions.

To effectively implement OSS systems as a public matter, Japan developed the following set of responses to each issue:

(a) Improper design → Setting structural standards and government procedures for approval and performance testing.

(b) Lack of monitoring of compliance of building standards → Setting a body of inspectors.

(c) Poor installation → Implementing systems for the registration and certification of private sanitation business and workers.

(d) Improper sludge management → Enactment of OSS systems (Johkasou Act) and implementing the obligation of the regular desludging.

(e) Unregulated sludging operators working under difficult conditions → Setting an approval system for desludging vendors.

(f) Improper treatment/disposal of on-site sludge → Developing sludge treatment facilities nationwide.

(g) Improper O&M → Enactment of the on-site system (in Japan the Johkasou Act) setting the owner's legal obligation for O&M, the owner's obligation of deploying a technical supervisor for a large OSS system ($\geq$501 population equivalent) and establishing the registration system for O&M vendors.

(h) Lack of human resources for maintenance works → Developing training, certification and examination systems for O&M technicians.

(i) Lack of awareness of OSS systems among owners and local governments → Establishment of a training institution for the professionals in the business related to OSS systems.

(j) Lack of accountability → Performing legal inspections.

(k) Poor O&M of large OSS systems of commercial users → Monitoring under the Water Pollution Control Laws (compliance with the effluent standard, measurement, report and inspection).

Source: adapted from Hachimoto (2021).

Table 2.3 Comparison among the models used to set the national sanitation plan.

Model	Advantages	Disadvantages
Fully centralised	• Generally, less costly to build and operate per m³ of processed water. • Functions well for urban areas. • Easier to operate by a specialised agency.	• May not be adapted to localised conditions, for instance when the ground has no slope or it is too hard to install sewers at an affordable cost. • Specialised agencies may be far from deprived regions' needs. • Rural areas may experience operation challenges. • Very low flexibility.
Decentralised	• Functions well for rural areas with disperse population. • May provide good solutions in urban areas with particular characteristics, for instance when the soil is highly permeable and any sewer leak can impact local water sources. • Highly flexible.	• Higher operation costs, and for extensive regions higher investment cost. • Demands variable professional skills.
Centralised/ decentralised	• If well designed and operated, it combines the advantages of both systems.	• Higher operation costs than the centralised model but lower than the decentralised option.

Source: partly, with information from AFBD, UNEP & GRID-Arendal (2020).

2.3.2 The institutional sanitation framework

The institutional framework is a set of formal governmental institutions, and their functioning, at the national, regional and local levels that under the national sanitation policy provide sanitation. The functions of all the institutions must be well defined and delimited, with clear mandates and procedures for cooperation. All these arrangements are to be reflected in the legal framework. Note that often the institutional frameworks for sanitation and for water supply are the same.

2.3.2.1 Participation of different sectors

To treat wastewater and implement a reuse project it is necessary to have permits from the ministries of health, the environment and the sector where reuse will take place. For instance, if the water and/or the sludge produced during treatment are to be used on agricultural land, permits from the agricultural ministry are needed. Thus, for the provision of sanitation services, in addition to the water sector, other government departments/ministries are involved, such as health, environment, building and construction, welfare, Indigenous people, agriculture and energy. Each one of these institutions deals with specific issues according to their competence and are led by their own ministries national policies. They can have different types of departments involved at the national, regional or local levels. In practice not all ministries operate in accordance with the institutional frameworks. In fact, mostly in the Global South, different ministers and ministries do not have the same power, political support and institutional capacity as each other. This depends on the sector and on the personalities and backgrounds of the ministers. Nevertheless, a robust institutional framework as an important step to achieve sustainable sanitation management is necessary (AFBD, UNEP & GRID-Arendal, 2020). The different institutions involved should have clear mandates with overall coordination mechanisms, expressed properly through the legal and institutional administrative frameworks ensuring that everybody covers their functions efficiently (Heller *et al.*, 2020). These functions are based on their sanitation policy and strategy definition, their coordination with the relevant actors; the planning, the implementation, the monitoring, the evaluation and the programmes' feedback, the definition and operation of their administrative procedures, the provision of inputs to elaborate/improve the legal and institutional frameworks and the development of human and institutional capacity. For some of the institutions, there will also be functions consisting of the operation and construction of infrastructure. The ministries contributing to the sanitation chain include:

- The Ministry/vice-minister/Commission/Department of Water (for planning, coordination and implementation),
- The Ministry of Health (to prevent negative effects on human health),
- The Ministry of Environment (to control water and soil pollution),
- The Ministry of Housing/urban development (to guide the users' requirements),
- The Ministry of the Interior (to coordinate the participation of different regional and local governments, and major stakeholders, such as the farmer's association and some civil group organisations),
- The Ministry of Agriculture (to decide when reclaimed water and sludge are to be used on productive soils),
- The Ministry of Indigenous and Vulnerable People (to ensure addressing the needs of all) and
- The Ministry of Economy (to control industrial wastewater pollution).

2.3.2.2 *National, regional and local levels*

The institutional arrangement for sanitation operates on three levels, namely: (i) national, (ii) provincial, regional or federative and (iii) local. Public sanitation tasks need to be carefully analysed so they are retained at the appropriate level (central, regional, basin or municipal levels/commune) according to the political, social and legal conditions. Figure 2.2 provides an example of each level, the participating institutions and their main functions.

National level. At the national level, the main function is to provide an enabling environment to implement sanitation. The national level needs to have a leadership role to define the national policy and to develop the entire

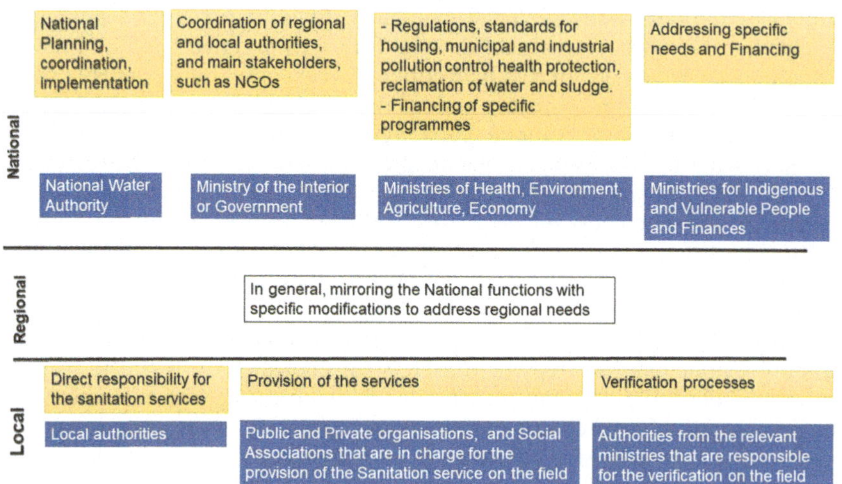

Figure 2.2 Sanitation service functions and responsibilities per level and organisation.

institutional and legal frameworks. Additionally, the national level needs to facilitate the financing mechanisms. Setting an appropriate national institutional framework is not an easy task; many models can be followed (Box 2.4). In several countries the management of water services is centralised in a single organisation, and this authority is supported by regional and local authorities.

Regional level. The regional level often mirrors the national structure. However, at this level, it is frequently decided whether private participation is allowed or not, and under what conditions. Also, at the regional level the water use is better understood. There are some countries where at the national level, water is part of a ministry (often environment or infrastructure) whereas at the regional level, water is a ministry in itself, with clear differences in the progress of the sector in the regions where water has a higher political relevance. The regional level also plays a relevant role in financing and setting practical regulations to promote sanitation (Box 2.5).

Local level. As sanitation is a service provided at the household level, the local level is key. Unfortunately, the local level frequently becomes the limiting factor in providing a good and reliable service. Hence, it is advisable to reflect during the design of the institutional arrangement on the relevance the local level has. Although, the local authorities should have a leading role coordinating implementation tasks, frequently they are understaffed and do not have a proper budget, and thus failures are observed. In addition, the local governments receive minimal guidance, information, allocated responsibilities and funds. This is especially critical in low-income rural areas. The complexity of an institutional setting for sanitation at the local level varies as the municipalities/villages/communities vary in size.

Box 2.4 Single national agency providing the framework for sanitation but with responsibilities differentiated at the national, regional and local levels

Mexico with 130 million inhabitants is the 14th economy of the world. The country is divided into 32 federative entities (regions) distributed along 1,964,375 km² (13th biggest country in the world). Mexico's topography and climate conditions are highly variable. In general, the northern part is arid or semi-arid whereas the southern part is humid. The country has 69 official languages. Of the total population, 79% live in cities whereas the rest are placed in highly dispersed rural communities (5.6 million people live in communities with less than 249 inhabitants). All these factors create conditions in which providing 'sanitation for all' is a daunting challenge. Following a centralised approach to manage water, Mexico has a single national water agency (CONAGUA). Since 2012, the HRtWS is a constitutional right. The national government must guarantee this right, and the National Water Law defines the conditions and support to be provided for its procurement. The local level has the responsibility of the implementation, the regional level supports and finances the related activities and the national level sets the National Policy, general procedures and standards, and provides technical advice and financing for the local and regional levels, depending on their own financial capabilities. At the regional and local levels, the provision of drinking water and sanitation services are tasked to the same organisation.

In 2021, the national sewerage coverage was of 95.2% and for the majority of the 1.5 million people still lacking this service, OSS systems are or will be provided. Nearly 67% of the domestic wastewater produced is treated (i.e., 145.3 m³/s). To reduce groundwater and surface water overextraction, 93% of the treated wastewater is reused, mostly for agricultural irrigation and industrial activities.

Source: with information from CONAGUA (2022).

2.3.2.3 Institutional stakeholders' and partners' involvement

All the stakeholders and partners along with the government must share the decision-making processes, each fulfilling their specified role. This is possible only if trusting relationships are built, in which information, benefits, tasks and risks are shared, and everybody plays their role and fulfils their responsibilities. Improvement is perhaps still necessary in clarifying and fulfilling the roles, responsibilities, rights and obligations of civil society (UN-Water, 2015a). Although participatory processes take longer, they reduce the risk of failure for the government when active participation of non-government parties is needed. This participation must be clearly acknowledged in the Sanitation Policy and in the institutional design. The institutions managing the sanitation services access and delivery include: (i) water/sanitation authorities, or municipalities, (ii) water/

Box 2.5 Example of an institutional arrangement for OSS systems at the regional level in Khulna District, Bangladesh and challenges observed at the local level (Khulna city)

At a national scale, Bangladesh has made a remarkable improvement in sanitation coverage by reducing open defaecation to 1% in 2015 from 34% in 1990, with an improved sanitation coverage of 61% in the same period. However, 28% of the population still use shared latrines and 10% use unimproved latrines.

Khulna district is located in the south-west of Bangladesh with a total population of 1.5 million and a population density of 32,859 persons/km². It is an important hub of trade and commerce. The district has 36 municipalities, 22 towns, 31 wards and 1,134 slums. The climate is generally humid during summer and temperate in winter with an average rainfall of 1,605 mm and temperatures ranging from 12.5 to 35.5°C. The institutional arrangement for delivering non-sewered sanitation systems and faecal sludge management (FSM) services in Khulna is vested in two government organisations: the Khulna City Corporation (KCC) and the Khulna Water Supply and Sewerage Authority (KWASA). Both are part of the Local Government Division of the Ministry of Local Government, Rural Development and Cooperatives (MoLGRD&C). The Conservancy Section of KCC is the main unit responsible for the provision of sanitation services, especially OSS facilities emptying services, street sweeping, surface drainage cleaning and solid waste management. KCC along with various NGOs and informal private service providers are major stakeholders responsible for the provision of OSS systems emptying services in many parts of the urban areas. KWASA is a corporatised utility with the mandate to provide water supply, drainage and sewerage services. KWASA activities are mainly financed from tariff revenues. Capital investments are usually financed by the central government.

The roles and responsibilities of the various actors involved in the broader FSM services are defined in the Institutional and Regulatory Framework for Faecal Sludge Management of 2017. This instrument clarifies the roles and responsibilities for FSM, re-affirms the ministries' lead role in policy making, the role of the city corporation and municipalities and the need for partnership with the KWASA, where relevant. The framework also provides the guidelines for the design of households and communal treatment facilities; specifies the role of private sector participation and identifies the need of the MoLGRD&C to set-up a unit within the KCC and Khulna's 36 municipalities for the delivery of FSM services with special reference to emptying of OSS infrastructure.

The MoLGRD&C sets priorities for the: (i) management of faecal sludge from septic tanks and pit latrines such that all sludge should be collected, transported and disposed, safely and regularly in an environmentally friendly manner; (ii) development of innovative technologies that are appropriate to local conditions for emptying, collection, treatment and safe disposal of faecal sludge; (iii) building of FSM and regulatory capacities of local governments in Khulna; (iv) implementation of relevant national sanitation policies and laws, as well as bylaws for timely, safely and regularly emptying septic tanks and pit latrines and (v) the provision of technical and business support to the private sector in sludge management, recycling and sale of compost or other byproducts (MoLGRD&C, 2017). The Bangladesh National Building Code (BNBC) is another critical legal instrument relevant to the improved management of OSS systems emptying and other associated technical operations and maintenance. This regulation makes provision for the formal approval of septic tank design and construction with appropriate technical specifications for controlled effluent discharge into the subsurface disposal field and/or seepage pits where public sewers are not available. This regulation prohibits the effluents discharge from septic tanks into open water sources and prescribes minimum septic tank capacity for different types of households. Also, the BNBC specifies the emptying frequency of a septic tank to be within a minimum of every 6 months and a maximum of once per year.

At the local level, there is Khulna city. It has a population of 31,883 people and relies fully on OSS technologies such as septic tanks, pour flush and pit latrines. Septic tanks are dominant in the city centre in multi-storey buildings, whereas peri-urban areas and low-income communities mainly utilise pit latrines. The city has one decentralised wastewater treatment plant designed to handle final treatment and disposal of faecal sludge from OSS systems. Despite the institutional framework described above, Khulna city faces challenges emptying its OSS systems. KWASA may require different management strategies to improve their capacity and competencies to improve service delivery for OSS.

Source: with information from Cookey *et al.* (2020).

sanitation service providers, (iii) water/sanitation services intermediaries and (iv) water/sanitation boards, and water/sanitation services committees (Figure 2.3). The institutional decision-making process comprises activities on sanitation, water reuse, wastewater, excreta and greywater management. At the national level, ministries and other stakeholders collaborate on the management of irrigation water, soil improvement and fertilisation in agricultural, forest, landscapes and protected areas, and in fishing activities in regards to public and environmental health protection and trade.

To note, at the local level, the institutions – especially if they serve small communities – are frequently the same for the provision of water and sanitation services. This is because the expertise required is similar; therefore having both services in one institution optimises both human and economic resources. For countries with Indigenous communities, the management of water services at

Figure 2.3 General institutional structure for delivering sanitation (and water) services in most countries. At all levels there is stakeholder participation through committees or commissions.

the local level follows their traditional rules using traditional organisations. A critical challenge to address is the integration of technical and scientific knowledge with local and indigenous knowledge to improve the selection, implementation and management of sanitation.

2.3.3 Legal sanitation framework

The purpose of this book is not to advise on how to produce a legal framework, as in most if not all cases, this already exists. Therefore, the aim is to provide recommendations on how to improve or reform it. The water legal framework should be produced to provide legal support for the objectives stated in the National Sanitation Policy and its institutional framework, not the other way around. However, because both the National Sanitation Policy and its institutional framework are often designed after a legal water framework has been set-up (in the best-case scenario a water law)[9] compromises are made when designing both. This is one of the reasons why the legal sanitation framework often needs improvement, notably to ensure that the national water law, water policy and sanitation policy are aligned to constitutional laws (Box 2.6). The legal sanitation framework comprises the National Water Law (when available) and all the articles coming from the Constitution and different national laws from other sectors, norms and standards that operate at the national, regional and local levels and are related to sanitation. These articles may come from the health, environment, building and construction, urban and Indigenous People

---

[9] To note that the national water law, the water policy and hence the sanitation policy and institution framework have to reflect and be aligned to constitutional laws.

Box 2.6 Basic sanitation policy in Brazil: legal and institutional framework

Basic sanitation services in Brazil have undergone a substantial process of institutional and legal transformations in the last five decades, since the establishment of the National Plan for Basic Sanitation (PLANSAB) in the 1970s, during the military dictatorship. During these last five decades, drinking water coverage increased from around 60 to 93% and the sewage network coverage went from 22 to 63% in urban areas; however these percentages hide the significant differences among the macro-regions and the rural areas of the country (MDS & SNS, 2021). The urban sewage service index varies from less than 10% in the north to over 70% in the southeast. These values do not include the number of septic tanks, which are very common in the north, as these are considered adequate by the PLANSAB. Additionally, nearly 80% of the sewage is treated before discharge, whereas the rest is discharged without treatment creating pollution problems (MDS & SNS, 2021).

To adapt to the changes of the National Sanitation Policy, the institutional and legal frameworks underwent several change phases. In 1971, a National Plan for Sanitation (PLANASA) was established with the Decree No. 949/1969. This plan was supported with an institutional setting comprising:

- Planning at the national level, undertaken by the Federal Service for Housing and Urban Planning (Serfhau) and, at the state level, diverse public Sanitation Enterprises (CESBs, State Basic Sanitation Companies in Portuguese),
- Implementation, undertaken by the municipalities acting as the title holder,
- Regulation and inspection of the tasks, undertaken by the Ministry of Interior (Minter) at the national level,
- Operation, undertaken by the CESBs with full responsibility at the state level and
- Financing, undertaken at the national level by the National Housing Bank, at the state level by the CESBs and at the local level by the municipalities. For financing, the users contributed too through their fees.

This institutional and legal setting allowed for the reduction of water deficits and increasing supply and sewerage coverage. A marked improvement, especially in cities, where the municipalities were the only institutions responsible for the services was observed. However, this model maintained a centralised national planning process and services provision was performed by state executive controlled enterprises.

In 2007, a new National Basic Sanitation Policy was developed (Law 11.445/2007) which was complemented, in 2013, with a PLANSAB.

This national plan encompassed regionalised goals that were set for the short, medium and long terms. Additionally, it did not impose the same institutional design everywhere; instead it allowed a mosaic of different solutions that could easily be harmonised with public and private participation and with regional and local needs. This framework also allowed innovative institutional design for basic sanitation such as public consortia, and allowed the title holder (the municipality), to delegate some tasks to third parties concerning the organisation, the regulation, the supervision and the provision of the service. A regulatory and inspection agency was created. The establishment of national guidelines and the possibility of providing contracts and concessions promoted the interest of states (regional level) to participate in sanitation tasks.

In 2020, a new legal sanitation framework was set with the Law no. 14.026/2020. The target set was to attain 90% of the sewage collection and treatment by 2033. The new law, instead of maintaining the state monopoly on the provision of the services, requires tendering processes in which private and public enterprises as well the State Basic Sanitation Companies (CEBs in Portuguese) need to compete to grant the service. This new model expanded the privatisation in the water sector. The National Water and Basic Sanitation Agency (ANA) was created to set, among other tasks, national regulatory guidelines that must be followed to be eligible for the allocation of federal resources. The Law 14.026/2020 also encourages the regionalisation of the services, such as PLANASA in the 1970s; however, now the regionalised provision is not carried out exclusively by CEBs but by any other public or private company. However, states continue to play a fundamental role, because they are responsible for establishing blocks of regionalisation, named regional basic sanitation units. The allocation of federal resources is conditioned to the structuring of regionalised provision and its respective adherence by holders of sanitation services. At the ANA, private companies, the Union, member states and municipalities, directly or through their agencies, public companies and other indirect public administration bodies are represented. Currently, the country has a National Agency (Agência Nacional de Águas e Saneamento Básico – ANA) and around 90 infranational (municipal, intermunicipal and state) agencies that operate under the regulation of sanitation sector.

In terms of the financing sources, the country moved from mostly public resources during PLANASA (Service Time Guarantee Fund – FGTS and State Water and Sewage Funds – FAEs), to an expansion of private participation. This has been a controversial aspect of the new law. It is well known that the privatisation process can pose a risk to guaranteeing HRtWS, because factors such as profit maximisation, the natural monopoly of services and power imbalance, typical of these processes, are not in line with human rights provision (Heller, 2020; Neves-Silva *et al.*, 2023). Another concern is that the new framework follows the

opposite course of the global trend, because many countries with priva-
tised services have returned to public sector management, in a process
called 're-municipalisation' or 'de-privatisation' (Kishimoto *et al.*, 2020;
Neves-Silva *et al.*, 2023).

Source: with information from Costa (2023), Cunha (2011), Werneck *et al.* (2020).

sectors, among others. In this context, the main role of the sanitation legal
framework is to define a clear and feasible mandate for the leading organisation
to implement coordination mechanisms for all sectors, institutions and
stakeholders. It is important that the legal framework sets or allows the setting
of clear procedures and definitions for the administrative procedures needed to
provide sanitation services.

There are three core issues that need to be clearly and fully developed in
the legal framework following the national sanitation policy selected. The first,
the institution mandates including coordination mechanisms, the second, the
models to involve private participation (enterprises, NGOs and international
organisations) and the third the criteria to protect health and environment from
the impacts of wastewater, sludge or faecal matter disposal and for safe reuse.

2.3.3.1 Mandates and coordination mechanisms

Regulatory mandates and functions are often more clearly defined for water
supply provision than for sanitation because the tasks to provide the service take
place before users and are mostly performed by the government. In contrast, for
sanitation, the activities are performed after the users and by an ample set of
public and private entities (companies – big or small –, NGOs and international
organisations). There is hence 'less freedom for definitions' resulting in mandates
and responsibilities which are less clearly stated. Furthermore, designing a well-
defined government legal framework is a challenge. This is simply because the
entire framework is prepared by different sectors. Although efforts are made to
align it across institutions and in accordance with the water law, laws and rules
are produced at different times and there are varying interpretations of each
sector's responsibilities and of the water law. This results in unclear mandates
and definitions for the tasks at the national, regional and local levels. This
presents implementation challenges. It is therefore recommended to appoint a
diverse group of government and non-government stakeholders and partners
to contribute to legislative and regulatory review and drafting, holding public
engagement sessions with different sectors, and requiring utilities to review
the associated administrative procedures to have a useful and user-friendly
sanitation legal framework. Having clear definitions of the institutional
roles and responsibilities prevents duplicating efforts and overlapping and
conflicting mandates (AFBD, UNEP & GRID-Arendal, 2020), notably at the
local level where sanitation is implemented. In this context, it is of the upmost
importance to clearly define which institution, and at which levels, will have the
responsibility for the sanitation service coordination (GWP, 2008). Investing

time in defining a proper sanitation legal framework is worthwhile. A holistic approach to policies, institutions and regulations is described by Eastern and Southern Africa Water and Sanitation (ESAWAS) Regulators Association (ESAWAS, 2023); this document identifies how stronger regulators can play an important role in improving sanitation for under-served urban residents just by properly adapting the legal framework.

2.3.3.2 Conditions to provide the sanitation service

For water management and service provision, some countries set legal specifications for private participation under the national water law (following general guidelines provided by national constitutions), whereas others have specifications and conditions dispersed in a diverse set of legal instruments (GWP, 2008). The legal framework defines the possibility and conditions for subcontracting and outsourcing. Usually, private companies participate under contracts following open bids in which the provision of the service is delegated to a varying extent. The contract is the instrument in which duties and rights of both parties are defined. The infrastructure, which may be built by the same or a different private company, remains always with the property government (OECD, 2015). The contract should be as comprehensive and clear as possible in all the client-provider instruments. When writing contracts, policy and decision makers need to consider: the conditions for private companies to operate the entire or part of the sanitation chain (Box 2.7), the indicators to

---

Box 2.7 Public–private water partnerships

Public–private partnerships (PPPs) are nowadays focused on the management of specific water activities, such as increasing energy efficiency. Most of the contracts are smaller in value and less complex than in the past, opening the way for new regional and local players and industries to meet these challenges. The participation is based on performance contracting, with payments against outputs. The development of water PPPs has become country specific, with governments keen on developing their own PPP schemes – usually with hybrid features which do not easily fit in the traditional classification lease/concession/build–operate–transfer (BOT). Five BOT schemes have been identified worldwide:

(1) *Build-operate-transfer (BOT) and design-build-operate.* They are used to build new infrastructure. The private financing is obtained, often using risk mitigation tools such as guarantees. These projects do not usually present the challenges of the private sector managing an existing public workforce or an interface with household customers, but they provide the benefits of private investment, expertise and technology and sustainable operations. They are particularly used for desalination and big wastewater treatment plants in many countries in the Global South (such

as the Middle East, China, Mexico and Brazil). There is a strong competition from a large and growing number of international and regional enterprises from countries in the Global South.

(2) *Performance-based contracts (PBCs).* These projects focus on results, with payments conditional to output achievement. Often the public remains responsible for running the day-to-day operations but benefits from private sector expertise in specific key areas. A substantial element of these contracts is knowledge transfer and capacity building of the utility workforce. The contracts cover an ample set of activities, ranging from reduction of non-revenue water and leakage management to increasing connectivity. A successful example from Ho Chi Minh City, Vietnam, where a PBC leakage contract resulted in saving water equivalent to the volume needed to serve an additional 500,000 people and saved 23,000 kwh/day of power.

(3) *Performance/output-based management contracts.* These are used either to outsource the management of a utility or to bring expertise to the public management. These contracts are common in the Middle East and North Africa (Algeria, Saudi Arabia, Oman), Latin America (Tegucigalpa in Honduras, Colon in Panama) and Africa (Democratic Republic of the Congo).

(4) *Small-scale private operators' contracts.* These contracts are becoming more commonplace in the Global South. Many donor-sponsored water or sanitation PPP projects for rural and peri-urban areas have been successfully implemented and scaled up, with new local operators emerging.

(5) *Large-scale private operators' contracts.* Several Global South countries have consolidated large national private water operators setting standards for reclaimed water and stabilised sludge and faecal matter. This has taken place in the Philippines (Manila Water, Maynilad), Brazil, Malaysia, Russia and also in some parts of Africa (e.g., SSE (Sénégalais des eaux, in French) in Senegal which became independent of Saur).

Source: partly, with information from WB (2023a, 2023b).

measure the efficiency and quality of the service, the procedures for estimating the operating costs and mechanisms to report to the government, the way in which financial risks will be managed, the transparency mechanisms for the society to verify the performance of public and private partners in the framework of the contract and the type and range of decisions in which the private industry can participate. Many of these aspects are part of the terms for the bid, which additionally need to specify when infrastructure is to be built and the acceptable technology or its characteristics to fit to the local context, particularly considering economic and social conditions.

According to WB (2023a, 2023b), today's market is radically different from the 1990s (dominated by the large concession model and private participation) as it is now oriented towards more focused projects. Nowadays, it has become clear that the water market is the one characterised by the need to make huge investments with slow financial returns, notably when compared to other public infrastructure (mines, energy, roads, for instance). It is noteworthy that the water supply and sanitation markets are completely different, as for the first, it is easier to set user tariffs to contribute or fully pay for the operation.

2.3.3.3 *Reclaimed water, stabilised sludge and faecal matter standards*

Reviewing effluent standards and setting standards for stabilised sludge and faecal matter is likely the main task of sanitation policy and decision makers. Adapting standards for the disposal of reclaimed water to the specific contexts of the receiving water bodies rather than using uniform or arbitrary standards for an entire country is the main and never-ending challenge. The standards for the reuse of reclaimed water for irrigation also need to be considered. For sludge and faecal matter, the disposal vs its reclamation are the two options to consider. Some remarks with respect to the standards and their compliance for these activities are provided below:

- For the fulfilment of standards, it is relevant to consider whether at the national and regional levels there is an appropriate institutional capacity to apply and enforce environmental regulations for the entire sanitation chain, that is, the control of water pollution, the disposal of reclaimed water into water bodies and/or soil, the options to reuse and recycle water and byproducts and disposal alternatives for stabilised sludge and faecal matter.
- The feasibility of compliance for the diverse social, technological and cultural conditions, as well as for the different population sizes.
- The gradual compliance of new standards, notably if major changes were made.
- The selection of a standard carries an implication of the technologies that can be used to fulfil it. It is therefore relevant to keep in mind that wastewater treatment schemes carry historical decisions. As the first problem observed with wastewater was the accumulation of wastes and sand in sewers and pumping stations, the preliminary treatment removes garbage and coarse-suspended solids (racks and degritters). With time, once the impact of discharging wastewater into rivers (humid climate regions) was observed, treatment options to remediate the biodegradable oxygen demand were introduced. After this, and particularly when reclaimed water is discharged to water bodies instead of to the soil, accelerated eutrophication problems were observed in lakes, dams and slow flow channels. Processes to remove nutrients were then considered. All these treatment steps are reflected in the conventional parameters and values used to set conventional effluent standards. When water is to be reused, these standards might not be

adequate[10]. Using different treatment goals opens the treatment options and configuration schemes can change.

2.3.3.4 *Reforms to the legal and institutional frameworks*

For functionaries the legal framework does not only set guidelines to be followed but also sets limits, which must be met to avoid being fined, banned from working in the government or even put in jail (Box 2.8). Thus, in many cases to promote paradigm shifts, develop innovative approaches or put in place

---

Box 2.8 New policy approach to implement sanitation in schools requires reforms to the legal and institutional frameworks

At the start of 2019, an assessment of public primary schools in Mexico revealed that notwithstanding the costly efforts made for decades to implement programmes to improve the schools' infrastructure, nearly 64% had no internet connection, 27% had no drinking water services, 17% had no proper sanitation facilities and 14% did not have energy. Furthermore, most of the available infrastructure at public schools was deficient and not adapted to the local needs or for disabled people. All the programmes used to address these issues had been implemented through centrally designed and financed programmes. These programmes were working thematically and with 'one solution fits all' to address the 'demands for which we have previously designed the solutions', and resulted in ineffective solutions which were not always accepted by the users.

Applying a new approach, in 2019 the President Andres Manuel Lopez Obrador decided to directly finance the student's parental association for them to define both their needs and solutions. This way, the programmes were adapted to their priorities and the administrative costs of running complex government programmes were minimised as the students' parental association is in charge of implementing the solution. The programme is called 'The School is Ours' (or 'La Escuela es Nuestra', in Spanish). Parents can choose what to do to address: (a) the extension of school opening hours; (b) the provision of meals to students; (c) acquiring educational equipment and (d) rehabilitating or building new infrastructure, which includes sanitation. The amount of funds depends on the number of students at the school, being of around 12,000 USD for 2–50 students; 14,500 USD for 51–150 students and 35,300 USD for more than 151 students.

---

[10] The reuse of treated water for agricultural irrigation is another option for disposal, notably in arid and semi arid regions. This option allows for higher organic matter (measured as biological oxygen demand, BOD) content, nitrogen and phosphorus in the water as these compounds improve the characteristics of the soil and increase crop productivity.

Initialising the implementation of this programme was very difficult as the legal and institutional frameworks (including the one for water) were created to centrally manage sectoral programmes, preventing granting money directly to the parental associations. Changes were implemented in schools so that our programme advanced fast and now the progress is on track to provide access to drinking water in 80% of the schools and provide electricity to 95% of the schools, adequate sanitation in 90% of the schools and hand-washing facilities in 75% of the schools by 2024.

Source: with information from SEP (2024)

new technologies, the legal and the institutional frameworks are barriers that need to be adapted. In fact, the legal framework must be periodically updated. These updates should consider the specific needs of a region, the current trends in the sector and the governmental policy changes.

2.4 CHALLENGES FOR REFORMING THE SANITATION POLICY

There are three groups of elements to consider when reforming the sanitation government:

(a) *Social*. Ensuring that 'All' are considered under the new programme and within an appropriate, non-discriminatory time framework. This means to develop and put in place strategies to address the needs for isolated communities (Indigenous, rural, displaced, etc.) for which it is difficult or impossible to apply national, territorial and local policies, and there is a need to develop dedicated policies within the national sanitation policy. Additionally, the effective and efficient involvement of stakeholders along the sanitation chain must be considered.

(b) *Political and financial*. Obtaining political support and sufficient funds to build, replace and operate sanitation systems including reversing inequities and addressing vulnerable population needs, demands special efforts that have to be reflected in the national policy. It is important to highlight that providing sanitation for all is costly, even if economically beneficial (Box 2.9) and permanently needed.

(c) *Technical*. To align the legal and institutional frameworks to new sanitation policies and to efficiently use the capacity of stakeholders that participate in the provision of sanitation.

The main challenge for the legal sanitation framework is to encompass the entire sanitation chain instead of only considering the tasks traditionally performed by the government. It is necessary to clarify regulatory responsibility gaps and overlapping areas along the sanitation chain. This idea is supported by references detailing further actions, including creating or strengthening the enabling environment for sanitation, as well as expanding the mandate of regulatory authorities to move beyond the 'conventional sewered sanitation'.

> ## Box 2.9 Cost of energy: one of the main barriers to providing sanitation
>
> In many middle- and low-income countries, water services are intermittent due to the high cost of electricity, which can represent around 65% of the operating costs. High energy costs are often the reason behind wastewater treatment plants being unused, even when the infrastructure is in place. This occurs despite the benefits of investing in water and sanitation largely outweighing the costs. An investment of 1 USD in water and sanitation services has a return in the range of 5–28 USD by saving in public health costs, encompassing reductions in health care expenditures.
>
> *Source*: with information from de França Doria *et al.* (2021), Hutton and Haller (2004).

2.5 NEW APPROACHES FOR SANITATION POLICIES

To meet present and future sanitation challenges, it is necessary to transform the infrastructure and the way in which sanitation is conceived and water managed. Policy and decision makers are or should be proactively moving from traditional conceptions, seeking new management approaches, partnerships and business models. As emphasised in the Dublin principles, water resource planning must be regional, but management and action must be local: 'Think globally, act locally!' (Seppala, 2002). This could be interpreted as 'be aware of new international and national approaches and based on your own context decide what could be an adequate solution for you'. Below are some ideas to consider:

- *Full incorporation of sanitation at the basin level*. Basin organisations can mobilise, support and promote cooperation to improve the institutional capacity to apply and enforce environmental regulations to support sanitation, water reuse and by-product reclamation practices. Coupling sanitation with water reuse to augment the available resources as part of the planning of hydrographic basins is to maximise the benefits from reclaiming wastewater, improve the efficiency of water management, better allocate resources involving stakeholders and protect water as a resource. Sanitation and water reuse integrated into river basin planning result in more sustainable and resilient systems, benefitting from additional sources of financing for implementation.
- *Citywide inclusive sanitation (CWIS)*. CWIS is an approach to urban sanitation where all members of the city have equitable access to adequate and affordable sanitation, without polluting the environment along the whole sanitation service chain. It aims at fostering the sanitation public service of cities to provide sanitation as a reliable, inclusive and sustainable service. The aim is to ensure everyone has access to safely managed sanitation, resulting from a range of solutions, each adapted to the specific context. Rather than focusing on building infrastructure,

it fosters an enabling environment for adequate sanitation. This is achieved by strengthening the design and implementation of core functions of public responsibility, accountability and resource planning and management systems. CWIS requires a diversity of appropriate technical solutions, combining both on-site and sewered solutions, in either centralised or decentralised systems, considering both resource recovery and water reuse. Political will, technical and managerial leadership and new and creative long-term funding options for sanitation are needed as well as institutional arrangements and regulations, with incentives for the O&M of the full sanitation service chain. Funding for non-infrastructure aspects of service delivery, such as capacity building, household engagement and outreach and sanitation marketing are required as well as complementary urban services, including water supply, drainage, greywater management and solid waste management, incorporated into sanitation planning. All these activities target specific unserved and underserved groups (Narayan, 2022; Narayan *et al.*, 2021; Schertenleib *et al.*, 2021).

- *Reconceptualising wastewater treatment plants.* The design and installation of treatment plants must be understood as facilities for the recovery of a valuable resource, with the application of circular economy principles. Water can be reclaimed for the irrigation of green areas, domestic and farmland agriculture; environmental protection; industrial purposes; energy production and even for human consumption. Sanitation facilities could therefore be perceived as a company, with the potential of generating products. This could help in reducing social rejection to local wastewater treatment plants and generate incomes for the sanitation service.

- *Rethink building structures and functioning design.* In cities it makes sense to integrate on-site water systems within buildings or communities. This way the specific interests and drivers for the use of water can be considered in the decision making of water management, including the sanitation aspect. Water can be used more efficiently to create new, local water supplies by utilising decentralised, on-site water treatment systems. These localised solutions create opportunities for shared responsibility over water systems between the government and communities. As these systems are deployed, oversight and management are critical for ensuring the protection of public health. Buildings can produce a variety of alternate sources of water including rainwater, stormwater, foundation drainage, greywater, wastewater (blackwater) and condensate. When collected and treated properly, these water sources can be used for non-potable applications such as toilet flushing, irrigation, and cooling towers. On-site water treatment systems embody the principle of fit-for-purpose treatment to the necessary level for the specific use. Moreover, these systems can transform the way water is managed in buildings. For example, on-site water systems can reduce potable water use by up to 45% in residential buildings, and up to 75% in commercial buildings (SFPUC, 2023).

- *Integrating green and grey infrastructure*. Service providers can deliver more cost-effective and resilient services by integrating green infrastructure into their plans. Green infrastructure should be evaluated from a technical, environmental, social and economic perspective. Wetlands can filter wastewater effluents and thereby reduce wastewater treatment requirements. Natural wetlands in rural areas can be adapted to provide sanitation (Rivera *et al.*, 1995).
- *'Smart cities' programmes*. Urban populations are increasing, despite the challenges that high density populations entail. This is why 'smart' city principles are popular. At a local level, water utilities could work closely with urban planners to review the end-of-pipe and centralised concept of sanitation and look for more sustainable ways to use water and apply resource recovery (carbon, nitrogen and phosphorus).
- *Institutionalising water reuse projects to close water loops, including sanitation*. It is necessary to institutionalise closing water loops, involving stakeholders from different sectors, to maximise the advantages of water reuse projects as part of the sanitation chain and as a water source (Box 2.10). Consistency in policies, approaches and standards can reduce the burden on technology vendors and designers. Furthermore, utilities, institutions and civil society have an important leadership role in actively promoting integrated water resource management and building an enabling environment for decentralised and integrated water solutions.

> **Box 2.10 Sanitation, water reuse and recycling; an innovative approach which requires cooperation from an ample set of sectors: The San Francisco On Site Water Reuse Programme**
>
> San Francisco, California, became the first municipality in the USA to adopt a groundbreaking programme that encouraged buildings to collect, treat and reuse water on-site to meet non-potable demands such as toilet flushing and irrigation. The programme was promoted by the San Francisco Public Utilities Commission (SFPUC). The San Francisco's Onsite Water Reuse Program established a streamlined process allowing alternate water sources, such as rainwater, stormwater, foundation drainage, greywater and wastewater, to be reused in commercial, mixed-use and residential buildings. Implemented by four city departments, the Onsite Water Reuse Programme is a successful example of investing in collaboration and eliminating barriers to using water more efficiently.
>
> The first step was to establish a city ordinance that clarified the roles and responsibilities of each city department: SFPUC, San Francisco Department of Public Health-Environmental Health (SFDPH-EH), San Francisco Department of Building Inspection (SFDBI) and San Francisco Public Works (SFPW). The Non-potable Water Ordinance helped smooth jurisdiction and authority conflicts and promoted inter-agency

cooperation. SFPUC developed the Onsite Water Reuse Programme Guidebook to assist projects with the permitting process. Although the programme began on a voluntary basis in 2012, the installation and operation of on-site water systems was made mandatory in 2015 for new development projects with a footprint of 23,226 m² or greater. In 2021, the threshold for new developments was lowered to 9290 m² or greater.

Scaling up on-site non-potable water systems across the USA. In 2014, the SFPUC, with support from the Water Research Foundation (WRF) and the Water Environment and Research Foundation (WERF), convened a nationwide meeting of public health agencies, water agencies and research institutions for an Innovation in Urban Water Systems conference. The conference's goal was to identify common challenges and discuss achievable solutions for a path towards widespread application of on-site non-potable water systems. One outcome of the meeting was the development of the Blueprint for On-site Water Systems: A Step-by-Step Guide for Developing a Local Programme to Manage Onsite Water Systems (available online at www.watereuse.org/nbrc). The blueprint instructs the communities interested in developing a local oversight programme to begin by convening a working group, and then establishing monitoring and reporting requirements while providing clear direction for project sponsors and developers.

The Innovation in Urban Water Systems conference also confirmed that the critical issue communities face in implementing on-site reuse is the development of water quality standards and monitoring strategies to ensure protection of public health. To address this challenge, the SFPUC convened a public health coalition to evaluate existing water quality standards for alternate water sources, develop recommendations for regulating on-site non-potable water systems and establish uniform practice among states. The coalition included public health agencies from several US states and was supported by WRF and WERF funding.

An Independent Advisory Panel, appointed by the National Water Research Institute, was established to provide technical advice on the management of health risks and on the monitoring needs.

In 2016, San Francisco partnered with the US Water Alliance to convene the National Blue Ribbon Commission for Onsite Non-potable Water Systems (NBRC). The NBRC includes representatives from public health agencies and water and wastewater utilities from 15 US states, the District of Columbia, US EPA, US Army and two Canadian cities – Toronto and Vancouver. The NBRC is supported by the WateReuse Association. The NBRC advocates for consistent policy frameworks across cities and states to help increase the adoption of on-site non-potable water systems.

Source: with information from Kehoe and Chang (2018).

doi: 10.2166/9781789064049_0047

Chapter 3

Policy and decision makers: the key to put 'sanitation for all' into practice

KEY MESSAGES

- In many low- and middle-income regions those who are responsible for providing sanitation are not necessarily trained, thus guidance is necessary.
- Sanitation provision entails challenges with no universal solutions, therefore the person responsible for sanitation requires knowledge, innovation and common sense. Technical, managerial and communication skills are also needed together with a strong spirit of service.
- To be innovative and efficient when implementing their tasks, sanitation policy and decision makers need to fully understand their duties, and have a grasp on key technical topics such as understanding the meaning of 'sanitation for all', the effects of the lack of sanitation, options to provide sanitation in different settings, water reuse in association with water rights, risks to sanitation workers, management of solid wastes in sanitation facilities, climate adaptation and mitigation in the context of sanitation.
- Policy and decision makers are accountable for their actions: this is why they play leading roles in sanitation activities.
- In the sanitation service chain, every link counts. Policy and decision makers need to understand who is doing what and the role governments need to play and how the national policy, the institutional and legal frameworks need to be shaped in their own context for the chain to be effective.
- Every country and every region has its own specific vulnerable groups. Achieving sanitation for all requires a good mapping of the social groups, and considering them in plans for sanitation, monitoring the progress and providing financing for the required activities.
- Innovative technology is needed all along the sanitation chain, policy and decision makers must support its production and introduction into the service.

> It is evident that those responsible for basic sanitation, wastewater treatment and water reuse projects need technical, managerial and communication skills together with a strong spirit of service; after all, managing the urine and faeces of others is not a pleasant task. Many people do not want to hear about this task, despite the fact that sanitation services are indispensable for society to function and as a public service it supports the fulfilment of several human rights.

3.1 BACKGROUND

Most countries are committed to the sustainable development goal SDG 6 'ensure availability and sustainable management of water and sanitation for all' by 2030 and to the United Nations Resolution recognising the human rights to water and sanitation (HRtWS). These engagements are often declared in national policies and country's visions; however, it is necessary to turn what is on paper into concrete and effective actions. This is the role of sanitation policy and decision makers. To achieve this, they *only* need to understand and manage national policies, institutions, financial and human resource situation, legal frameworks and individual and collective knowledge, and to reach consensus with all stakeholders for them to be part of and assist in the implementation of projects.

This chapter will explain the tasks expected from policy and decision makers to achieve sanitation. Policy and decision makers do not have the same responsibilities. Policy makers are those defining objectives to achieve and provide general procedures, whereas decision makers must put into practice what the former designed. Briefly, policy makers define the road for decision makers to follow. However, in daily life, and especially at the lower government levels, policy and decision makers are the same person. Therefore, the tasks for both will be jointly cited throughout the text, unless clearly stated that the task is for only one of these two types of functionaries.

3.2 TASKS AND CHARACTERISTICS REQUIRED FROM SANITATION POLICY AND DECISION MAKERS

3.2.1 Understanding societal benefits from the provision of sanitation

Before developing policies, implementing projects, undertaking awareness raising, communication and education programmes on sanitation, it is important for those responsible for the provision of sanitation to understand and document the impact of its activities on the well-being of the public. This includes the improvement in hygiene and human health and the contributions to improve education, gender equity, economic conditions and the protection of the environment (see Chapter 4). Unfortunately, many people responsible for sanitation are not aware of its wider benefits and rarely have data to prove these (Box 3.1).

Box 3.1 Decision and policy makers need to know and use information on the positive daily life impacts of sanitation: case study in Lima, Perú

A study to compare the health risks associated with the consumption of crops irrigated with three different water sources with distinct quality parameters in the Lima market was carried out in Peru. The three types of irrigating water sources were: (i) raw domestic wastewater from San Martin and Callao; (ii) treated wastewater from San Juan and (iii) 'clean water' from the Cieneguilla River. In all cases, both the groups consuming the crops and the local policy and decision makers were unaware of the microbiological quality of the irrigating water sources and the associated effects.

The evaluation of the quality of water and crops and the assessment of risks showed (Figure 3B.1) that the group with a higher exposure to pathogens were, as expected, the group consuming vegetables irrigated with untreated wastewater. The risks for those consuming crops irrigated

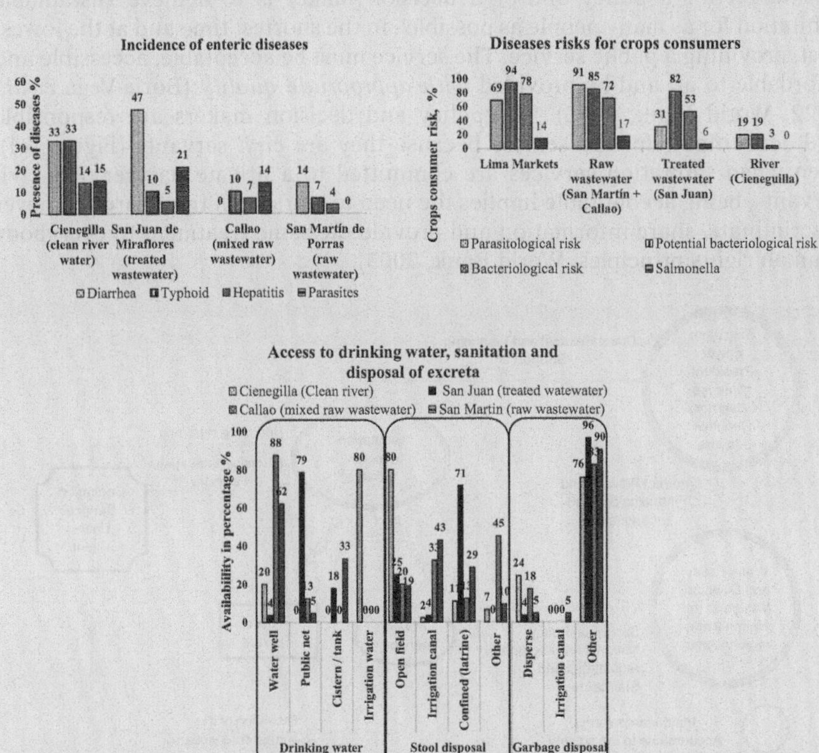

Figure 3B.1 Enteric diseases in farmers, and risks for consumers of crops from water reuse areas with different conditions of drinking water, sanitation and stool disposal.

with treated water and with clean river water were significantly less and very much lower, respectively. Regardless of the type of water used for irrigation, the microbial and parasitological risks due to the consumption of vegetables commercialised in the Lima market were very high (69% for parasites and 78% for pathogenic bacteria, see graphs), almost same as the risks associated with the consumption of crops irrigated with non-treated wastewater (91 and 72%, respectively). These risks originated from the water used for washing of crops, vegetable manipulation during transport, cooling and commercialisation. Acquiring all this type of information and communicating it to the public helps to understand the relevance of sanitation and controlling water pollution in daily life.

Source: with information from Castro de Esparza *et al.* (1990).

3.2.2 Accountability of civil servants and service providers

The target for a policy and/or a decision maker is to achieve sustainable sanitation for as many people as possible[1], in the shortest time and at the lowest cost, providing a public service. The service must be acceptable, accessible and affordable to all and be provided *with appropriate quality* (Borja-Vega *et al.*, 2022; World Bank, 2003). The policy and decision makers are responsible and accountable for the service because they are *civil* servants (Figure 3.1), even when sanitation services are committed to a private partner. For civil servants, being accountable implies the need to work with transparency, never discriminate, share information and provide the same treatment to everybody (human rights principles, World Bank, 2003).

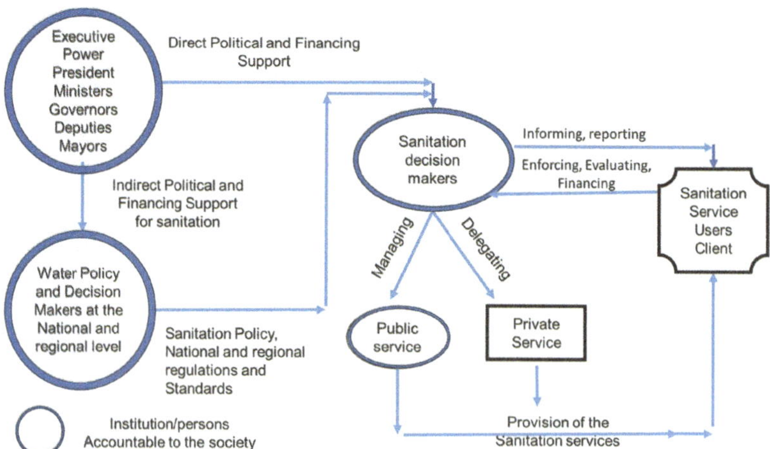

Figure 3.1 Accountability for civil servants when the government is the only one providing the sanitation services.

---

[1] Keeping in mind that the overall goal is to achieve 'sanitation *for all*'.

The fact that the service has to be provided for the entire sanitation chain[2] means that non-government stakeholders are (or should be, in case the legal framework does not yet consider it) also accountable when there is a client–provider relationship. This accountability remains in the field of the commercial services and does not undermine the government accountability. The World Bank (2003) introduced a model in their World Development Report 2004 to evaluate water, sanitation and hygiene (WASH) service delivery establishing relationships between power and accountability[3], for both commercial (client/provider) and public (government/citizen) with the different actors involved in this service chain, namely citizens/clients, politicians and policymakers, organisational providers, frontline providers and external financing agencies. On the basis of these ideas, Figure 3.2 presents different pathways for the two types of accountabilities. The traditional public one, involving the government/citizen relationships, the commercial one (client/provider) and a combination of both when the government is subcontracting/delegating the service. Two key relationships – voice[4] and

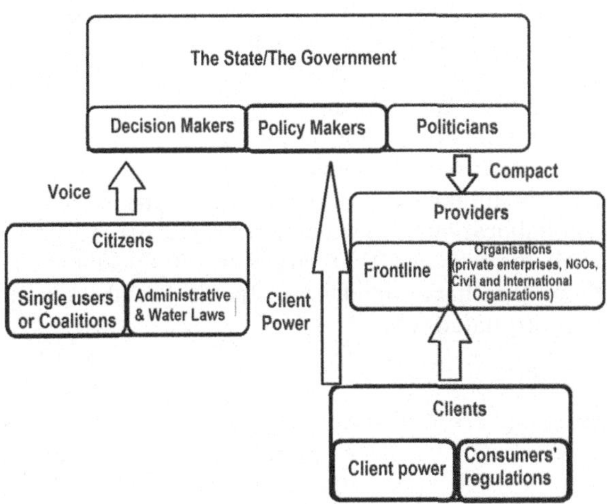

Figure 3.2 Relationships of power and routes for accountability for both civil servants and different stakeholders participating in the provision of the services as part of the sanitation chain approach. (*source*: adapted from Borja-Vega *et al.*, 2022; World Bank, 2003).

---

[2] The 'sanitation chain': involves the experience of the user, the excreta and wastewater collection methods and processes, the transportation and treatment of water and wastes, and the water reuse or disposal.

[3] Accountability: relationship among actors that has five features: delegation, finance, performance, information about performance and enforceability (World Bank, 2003).

[4] Voice: avenue connecting citizens and politicians and comprises many formal and informal processes, including voting and electoral politics, lobbying and propaganda, patronage and clientelism, media activities, access to information, and so on. Citizens delegate to politicians the functions of serving their interests. Politicians perform by providing services, such as law and order to communities (Borja-Vega *et al.*, 2022).

compact[5] – constitute the main citizen control mechanism in this route (Borja-Vega *et al.*, 2022; World Bank, 2003).

3.2.3 Understanding institutions' role

A sustainable sanitation service is an industrial chain in nature, and every link counts. As mentioned in Chapter 2, several institutions from different sectors and government levels participate in it. This is a challenge for decision makers when implementing projects, as they need to understand who is doing what and if a good coordination is set-up for all procedures, not only on paper but also in practice. Firstly, a clear institutional framework is necessary, as different institutions may show a varying degree of efficiency in implementing their given mandate. Furthermore, politics play a role as in many cases depending on who is the head of a sector or institution, the leading roles change even beyond their mandates. This impacts the result and the ways in which decision makers will have to work at the local level. Additionally, it is important to understand the financing each institution receives as it defines the actual role they can play. There is no 'handbook' on the road to follow for decision makers, but good knowledge and understanding on the role each institution involved effectively plays *and its real performance*, certainly helps.

3.2.4 Understanding the three dimensions to act

Decision and policy makers will not only have to 'swim' across the legal and institutional frameworks to implement a national sanitation policy but will also have to act only in the intersection of three dimensions (legal and institutional framework and collaboration with stakeholders) (Figure 3.3). According to the World Bank (2003), the countries that have successfully advanced on sanitation coverage are those that have managed to tackle the above-mentioned three dimensions in a coordinated way.

3.2.5 Being a team player

Policy and decision makers for sanitation work in different institutions that are placed at the legislative or executive powers from the national, regional and local levels (Figure 3.4). There are also other relevant actors from universities, research centres, private and public non-governmental organisations and international organisations. A positive aspect of being a set of policy and decision makers aiming to achieve the same target is that one is not alone. But, of course, everyone must be able to work as a team.

Regardless of the level in which decision makers must act, all the activities of the sanitation chain are relevant. This is true for big sewerage systems, wastewater treatment plants (WWTPs) and treated water disposal or reuse systems both in urban and rural areas. As such, part of the decision-making process must be delegated to lower administrative levels: provincial, municipal or district authorities, Indigenous communities and so on. Possibly, the most

---

[5] Compact: the broad, long-term relationship of accountability connecting policymakers to organisational providers. This is usually not as specific or legally enforceable as a contract. But an explicit, verifiable contract can be one form of a compact (World Bank, 2003).

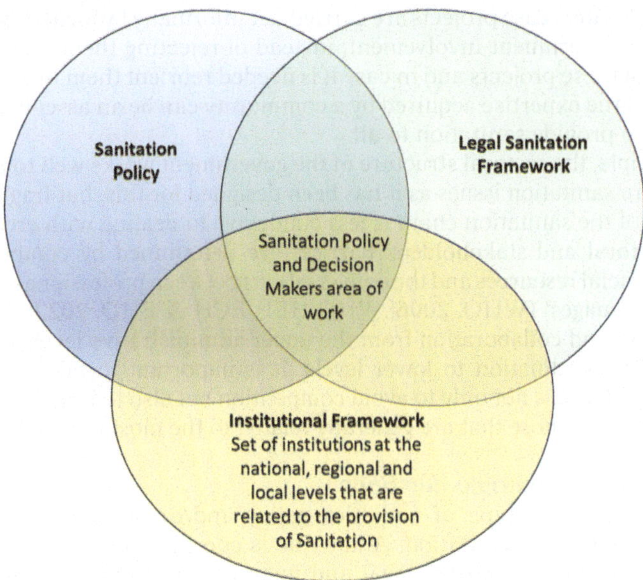

Figure 3.3 Range in which sanitation and policy decision makers perform their tasks.

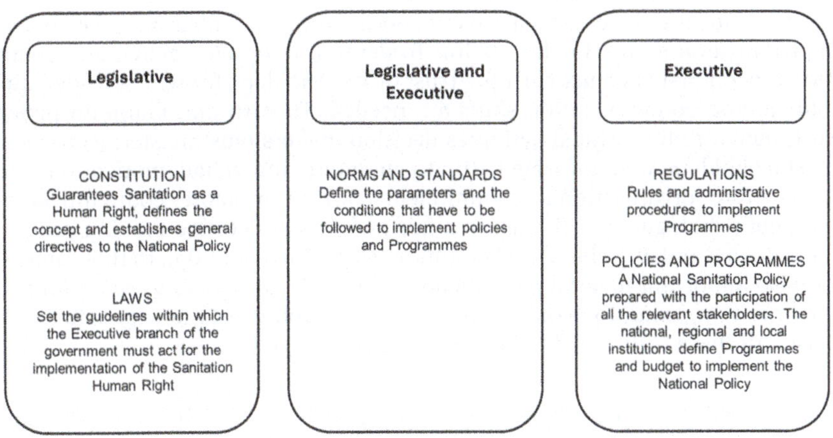

Figure 3.4 Responsibilities/functions of the executive and legislative powers concerning the sanitation government structure.

complex context to provide sanitation and ensure the success of the chain service is the one corresponding to small communities in isolated areas. Being a team player implies working alongside the other government institutions and collaborating on their projects. It also means working with all the stakeholders and partners, not only with those from your own project but also with those from other institutions and stakeholders. In many cases, small-scale sewage, excreta and water reuse projects are carried out informally by local communities without the government involvement; instead of rejecting them it is necessary to be part of these projects and in case it is needed reorient them in a diplomatic manner. All the expertise acquired by a community can be an asset in achieving the target to provide sanitation to all.

In principle, the sectoral structure of the government works well to effectively address core sanitation issues as it has been designed for this, but fragmentation of the rest of the sanitation chain is less conducive to dealing with cross-cutting issues. Sectoral and stakeholders' barriers are determined by competition for limited financial resources and the conversations between professionals 'speaking different languages' (WHO, 2006; WHO-HEP, ECH & EHD, 2022). Promoting coordination and collaboration from the upper administrative levels is essential to permeate coordination to lower levels. It is important to take coordinated decisions at all levels not only to avoid competition but also to be able to advance in small projects, those that are generally related to the most vulnerable people.

3.2.6 Performing multiple functions

Table 3.1 shows the type of activities policy and/or decision makers must undertake to achieve sanitation. A key role is coordination, and this must be clearly mandated through the legal and institutional frameworks to a single institution at the national, regional and local levels for specific tasks.

3.2.7 Reinforcing or acquiring specific skills

Governments are the critical actors in coordinating and integrating behaviour change in sanitation (Hartley, 2006). To develop desirable behaviours, policy and decision makers must earn the trust of stakeholders, for which leadership, political and communication skills are needed. To earn and maintain public trust, there are five critical activities decision makers must master: (i) to keep all stakeholders well informed, (ii) to maintain individual motivation and demonstrate organisational commitment, (iii) to promote communication and public dialogue, (iv) to ensure a fair and robust decision-making process and outcome and (v) to build and maintain trust (Hartley, 2006). In this context, it is also important to act diplomatically, *what is being said is as important as how it is being said*, to avoid negative reactions and to add new people to the cause. All these skills need to be and can be developed.

3.3 IMPORTANT TOPICS FOR THE UNDERSTANDING OF SANITATION POLICY AND DECISION MAKERS

On the basis of our experience, the following is a brief description of some topics that are indispensable for policy and decision makers to be well acquainted with (further technical reading is advisable, here we are only giving an overview). In many middle- and low-income countries and regions, the people who oversee sanitation do not necessarily have the essential background or experience. This does not hamper the possibility for them to obtain effective results in achieving sanitation targets. To shorten their learning period, a list of core subjects that they need to master with time is provided.

Table 3.1 Activities that policy and decision makers will have to undertake to provide sustainable sanitation services.

Activity	Description
Development of policy and strategy	The definition of goals and objectives along with the acceptable procedures to reach them.
Development of regulations, norms and procedures	Consists of setting standards and establishing rights and obligations, accompanied by procedures to ensure accountability. It covers the elaboration of formal legal mechanisms, enforcement processes and ensuring that institutions have a clear and useful mandate along with administrative mechanisms to perform their duties.
Development of programmes	The definition of concrete actions and processes to put the national sanitation policy in practice following the regulation framework. It includes planning based on the analysis of the current situation using available or collected data, followed by the formulation of actionable plans and estimation of cost for a specific period.
Management	The combination of the organisational, managerial and institutional arrangements at national and sub-national levels for the functioning of an organisation. In the sanitation service provision, this entails the definition of service delivery, model – who owns, who invests, who develops and who operates the infrastructure, who supervises and provides technical support – and the relationship among all these actors, with the users.
Monitoring and evaluation	The systematic processes of collecting, analysing, evaluating and using data to track performance and inform the planning and decision-making processes.
Preparedness	Refers to the arrangements, capacities and knowledge developed by governments, response organisation, external agencies, communities and individuals to anticipate and plan, to be able to mitigate and respond effectively to the impact of potential or current shocks and stresses.
Institutional and personal capacity development	Capacity development refers to the process by which organisations, society and individuals systematically stimulate, develop, strengthen and maintain their abilities over time to set and develop their goals and objectives to be able to manage sanitation services. Personal learning includes formal and informal processes, whereby stakeholders exchange good practice and information and use the newly acquired knowledge in managerial decisions to adapt and improve policies and programmes.
Coordination	Coordination comprises processes, mechanisms, instruments and platforms that promote and ensure multi-level, multi-sectoral and multi-stakeholder cooperation among all actors – relevant ministries and departments of central, regional and local governments, civil society, academia, external support agencies and the private sector. It entails information sharing.

Source: Elaborated partly, with information from Heller *et al.* (2020) and Jiménez *et al.* (2020).

> ## Box 3.2 'Clean means healthy'
>
> In Mexico City due to the swine (H1N1) flu outbreak in May 2009, restaurants were closed and public transport was stopped. This represented an economic loss of 144 and 35.2 million USD, respectively, in just 10 days. To allow the city to return to normal conditions, health experts advised constant hand washing and the disinfection of school toilets. This policy brought to the light that 200 public schools had no water, 195 had malfunctioning toilets and 90 had no facilities at all. Before the swine flu epidemic, politicians had not understood the link among water–sanitation–health and, as a consequence, had not addressed this problem, despite parents' associations having repeatedly requested proper water services at schools. The president of one parents' association commented on the news that, in contrast to most Mexicans, he believed that the swine flu had been a blessing for the city as it was the only way to ensure proper sanitation facilities at schools. The Mexico City government invested 56 million USD on the school programme 'Clean means healthy', a third of the economic losses mentioned above. When the COVID epidemic arrived, two decades later, the same situation was observed in schools but this time in the rest of the country.
>
> *Source*: with information from Jiménez-Cisneros (2011).

3.3.1 Effects of the lack of sanitation

The two main *direct* effects are those on health (Box 3.2) and on the environment.

3.3.1.1 Effects on public health

Basic sanitation and well-functioning wastewater management infrastructure makes it possible to prevent infectious diseases such as diarrhoea, intestinal parasites such as ascariasis[6], schistosomiasis[7] and trachoma[8]. Infectious diseases affect millions of people, especially children. Lack of sanitation is also a cause for vector[9] diseases that are originated by organisms (such as flies) that develop in stagnant parts of drainage systems, treatment lagoons or stored wastewater and that mechanically transfer pathogens to people or to their food. The biological risks depend on the epidemiological characteristics of the localities or areas. In addition to pathogens, the on-site sanitation (OSS) systems and non-sewered sanitation (NSS) systems contain organic matter, nutrients and a diverse set of chemical contaminants. Chemical contamination depends on industrial activities, and on the domestic use of cleaning, hygiene

---

[6] Roundworms.

[7] Also known as snail fever, bilharzia, and Katayama fever.

[8] Trachoma is an infectious disease caused by a Chlamydia bacterium. The infection causes a roughening of the inner surface of the eyelids, causing pain and eventually blindness which can be permanent.

[9] Vectors produce endemic and epidemic diseases such as malaria, dengue, chikungunya, zika, as well as the proliferation of rodents and crawling insects, which need to be controlled.

and personal care products. Biological and chemical wastewater pollutants and effects can be consulted in specialised literature (some suggested references are Jiménez, 2007; Jiménez-Cisneros, 2011; Nataraj, 2022; Schuster-Wallace *et al.*, 2014; WHO, 2006). The sanitation policy and decision makers need to be vigilant of emerging contagious diseases that have the potential to be transmitted through human wastes. This vigilance demands close and cautious follow-up as little is known on pathways for transmission (Box 3.3).

Box 3.3 COVID

The COVID-19 pandemic was caused by the severe acute respiratory syndrome coronavirus 2 (SARS-CoV-2) that spread worldwide, with over 170 million confirmed infections and around 3.7 million confirmed deaths as of 7 June 2021 (WHO, 2021). The infection is primarily transmitted from one person to another by contact and respiratory droplets that are expelled orally. A susceptible person may inhale these droplets from the air or, less frequently, the droplets may land on surfaces (such as tables and door handles) that people touch and subsequently transfer the virus to their faces through hands (Tang *et al.*, 2021). For these reasons, the measures to prevent the transmission were to maintain physical distance between people, adequately ventilate indoor spaces and wash hands with water and soap, which dissolves the lipid envelope of the virus.

Before knowing the pathway for infection, concerns were raised on the possibility that the infection may also be spread by faecal–oral route. We know now this is very unlikely to occur, but this required research and time. SARS-CoV was found in sewage wastewater and later it was known that it remains active from 2 days at 20°C to 14 days at 4°C (Wang *et al.*, 2005). During the pandemic, RNA from SARS-CoV-2 was detected in the faeces of patients with COVID-19, but it was mostly not possible to isolate the active virus, suggesting that the faecal route was unlikely to be the main source of infection, if at all infectious (De França Doria *et al.*, 2021; WHO, 2020). The SARS-CoV-2 was found to be quite sensitive to halogen disinfectants and ultraviolet disinfection (Tang *et al.*, 2021). For these reasons, treated wastewater and more importantly drinking water are unlikely to pose a significant risk if disinfected (De França Doria *et al.*, 2021).

The World Health Organisation (WHO) issued a technical brief for the prevention of COVID-19, targeting water and sanitation practitioners and providers (WHO, 2020). This document highlights many co-benefits that can be realised by safely managing water and sanitation services and applying good hygiene practices (De França Doria *et al.*, 2021). Besides the well-known negative impacts of COVID, for the sanitation sector it provided knowledge on the need to be prepared, by first gathering information, to be able to act according to the specific situation. Further, wastewater testing provides an opportunity to obtain advanced warning of diseases in a community. While known previously, this has become a mainstream tool in disease surveillance since COVID (Pelling *et al.*, 2021).

When untreated wastewater is discharged into the environment in areas with water scarcity, it often leads to unplanned reuse mostly to irrigate crops. High morbidity and mortality rates from gastroenteritis, dysentery and helminthiasis are observed when people consume such crops without proper disinfection (WHO, 2006; Dickin *et al.* (2016)). Those consuming the crops are not the only ones affected, as there are other social groups that are exposed, such as:

(i)　Operators of sanitation systems who manage the collection and treatment of wastewater and OSS system content.
(ii)　Farmers handling wastewater and polluted crops.
(iii)　Local community living near the irrigated plots or downstream of the canals conveying the polluted water.

3.3.1.2 *Effects on the environment*

Globally, 80% of all municipal and industrial wastewater is released into the environment without any prior treatment, resulting in an increasing deterioration of overall water quality with detrimental impacts on ecosystems and human health (WWAP, 2017). The most frequent challenge in water quality is usually associated with the load of pathogens, but no less important is the nutrient load which causes eutrophication[10].

To avoid affecting flora and fauna or to reuse reclaimed water it is important to treat it before disposal or the next use. The type of treatment must respond to the resilience or acceptance capacities of water bodies receiving the treated water or the characteristics needed for the next use, which can be agricultural irrigation (FAO, 2017) or an industrial or municipal reuse.

Emerging contaminants, including human and veterinary pharmaceuticals, personal care products, illicit drugs and naturally excreted hormones are ubiquitously detected in wastewaters, as well as surface and ground waters due to the incomplete removal of these pollutants in WWTPs and the direct discharge of wastewater to the environment (Branchet *et al.*, 2021; Gaw *et al.*, 2014; Mezzelani *et al.*, 2018; Mo *et al.*, 2022; Peña-Guzmán *et al.*, 2019; Świacka *et al.*, 2022). The negative effects of these pollutants on aquatic life have been widely documented, as well as the fact that antibiotics may also act as co-selection drivers in the development of antimicrobial resistance (Becerra-Castro *et al.*, 2015; Cen *et al.*, 2020; Marti *et al.*, 2014; Ventola, 2015).

Of note, when treated wastewater is discharged into the soil it acts as an additional barrier/tampon for water bodies as it biodegrades several times more organic matter than water bodies and reclaims nutrients to grow plants, while improving the quality of water (Jiménez-Cisneros, 1995) but some negative impacts can be observed in soil biota and potentially plants, particularly from emerging contaminants (Garduño-Jiménez *et al.*, 2023; Garduño-Jiménez & Carter, 2023; Gomes *et al.*, 2017).

---

[10] Eutrophication is a proliferation of plants – such as macrophytes, seaweed, cyanobacteria and algae – in lakes, low flow rivers and channels and even in the sea. It is caused due to a high nitrogen and phosphorus content in water (Chorus & Welker, 2021; Jimenez-Cisneros, 2001).

3.3.2 Basic sanitation[11] and wastewater treatment systems

In the Global South, there is a clear difference between the services for basic sanitation[12] and for wastewater management,[13] the last considered as the 'traditional' way to provide sanitation in the Global North. Basic sanitation is used where no sewerage is available such as in rural areas with dispersed population or low-income urban areas. In most middle- and low-income regions, the coverage for basic sanitation (notably the insufficient coverage of faecal sludge management) lags behind the coverage of wastewater treatment (which might not be very high either).

Households with NSS or OSS systems, most in need of faecal sludge management services, are often located in densely populated urban settings or rural areas. However, a quarter of urban sanitation policies or plans do not address faecal sludge management (emptying, transport, treatment and end use or disposal). Governments need to recognise the important role that the sound management of faecal sludge plays in achieving national sanitation targets and the SDGs, addressing this issue in sanitation safety policies and plans by providing them with adequate resources for their implementation (UN-ESCAP, UN-HABITAT & AIT, 2015; WHO & UNICEF, 2021). This is why it is important to keep in mind that for sanitation the service is provided after the users, and it consists of a chain of services.

In places where sewers are available, the population frequently has access to drinking water services at home, that may or may not be functioning continuously. Sewers, besides conveying household's wastewater, often convey rain water that is collected from streets together with non-controlled dumped solid wastes or even sediments coming from soil erosion. The sewers may end up in WWTPs, which in turn may or may not be well functioning. Sewers and WWTPs when available end in pipelines discharging the treated, partially treated or untreated wastewater into water bodies, the soil or irrigation channels.

---

[11] The definition for basic sanitation used here does not follow the JMP sanitation ladder reference, as we refer here to how it is used in the Global South. The JMP produced its definition to provide a 'continuum' in the way they used to conceive and monitor the practice (WHO-UNICEF JMP, 2016).

[12] Basic sanitation is understood here as what WHO (2018) Guidelines on Sanitation and Health, defines as sanitation, which is 'the access to and use of facilities and services for the safe disposal of human urine and faeces. A safe sanitation system is a system designed and used to separate human excreta from human contact at all steps of the sanitation service chain from toilet capture and containment through emptying, transport, treatment (in-situ or off-site) and final disposal or end use'. In this definition, the transport of the content of OSS and NSS through trucks, the management of wastewater and the systems to dispose of or to reclaim treated water and by-products is not considered.

[13] Wastewater management systems are the systems used to collect wastewater and OSS and NSS content, to provide its treatment and either used for its disposal or reuse.

3.3.3 Sanitation vs water reuse (circular economy)

Wastewater is conceived as a resource in the context of a circular economy (WWAP, 2017). This change in perspective is required to consider the sustainable management of water and eventually the existence of incentives to promote investments in the management of wastewater, where the reuse and recovery of the resource and associated compounds are considered. For this change, society must be aware of some facts, such as:

- water resources, and other resources such as phosphorus, are limited,
- water is a resource that cannot be destroyed and is naturally 'recycled' through the water cycle,
- directly or indirectly, society is contributing to environmental pollution through the use of water,
- several water pollutants can be used as raw material for economic activities and
- 'wastewater' is 99.9% or more just water.

Wastewater is still seen as a substance that must be treated and disposed of, rather than a resource. This produces a lack of political will to develop policies and regulations that support and encourage the reuse of water and the recovery of the resources contained in it. One way to promote resource recovery, including water, is to promote social responsibility. This has happened in many sectors, for instance in fashion industry where it is trendy to commercialise used clothes, or in solid wastes, where it is not well seen to perform waste separation at home to promote recycling. Circular economy should embrace the water sector. Conceptually, the circular economy is a model of production and consumption which involves sharing, leasing, reusing, repairing, refurbishing and recycling existing materials and products as long as possible. In this way, the life cycle of products is extended. It is an 'economy that preserves the added value of products for as long as possible and practically eliminates waste' (European Commission, 2010). In the case of water, this approach makes sense.

The economic factor approach has the advantage of influencing the perception of the benefits that the use of wastewater can generate (Doménech, 2011). The inadequate valuation of water also leads to the generation of erroneous rates for water resources and services, which discourages the development of resource recovery projects. For example, if industries pay a very low fee to extract fresh water, they have little incentive to pay for reclaimed water, unless there is a recurring water shortage (WHO/CED/PHE/WSH, 2018). In this new mind frame, sanitation would be part of a system in which it makes sense to financially support this service to promote water reuse and protect water sources (Box 3.4).

3.3.3.1 *Water reuse for agricultural irrigation*

By far, the most frequent reuse of water is for crop irrigation, simply because worldwide agricultural irrigation demands the most water (around 70% globally). Water reuse for irrigation happens with treated and with non-treated wastewater. In water-scarce areas it is the lack of sanitation that provokes the non-controlled reuse of water. It does not necessarily happen because the

Box 3.4 Tenorio project: a success story of sustainable development in San Luis Potosí, Mexico

Industrial and economic development in the city of San Luis Potosí (with nearly half a million inhabitants in 1990) has always been related to the availability of water. Since 1961, the extraction of water from the two main aquifers was restricted and farmers had to use untreated wastewater for irrigation. At the end of the 1990s, the government decided to implement a Comprehensive Water Sanitation and Reuse Plan in which reclaimed water to be used for irrigation and other non-potable uses that were still using groundwater. The combination of both sanitation and reuse, notably for industrial purposes, made it economically feasible to treat wastewater. Under this plan, wastewater treatment was increased from nearly 0 to 70% and all treated water reused by 2004. The reuse programme for industrial purposes made the plan economically viable while it increased water resources' availability. The project not only had economic benefits, but also a positive impact on the local community, in terms of improving public health and the environment.

The project was achieved with a single WWTP with a capacity of 90,720 m³/day. The treatment consists of an advanced primary treatment (Jiménez & Chávez, 1998). Of the effluents, 43% is sold to a power plant ('Villa de Reyes') where it receives an additional treatment consisting of activated sludge with nutrient removal, lime softening, sand filtration and ion exchange for silica and hardness removal processes to fulfil the water-cooling standards. The treated water is used to replace the water that was being extracted from the overexploited aquifer. The rest of the effluent, 57%, is treated in a wetland before using it to irrigate 500 hectares of forage crops. To distribute treated water among farmers a complex irrigation system (several pumping stations, an irrigation network, 39 km of pipelines and a surge tank to accommodate hourly industrial demand) was built. The reclaimed water complies with the Mexican standard for irrigation (NOM-001-SEMARNAT, 2006), with the government having the responsibility for its treatment and water quality.

The WWTP, the irrigation system and the 59 km of sewerage pipes required a total investment of 67 million USD in May 2004. The project was built following a build–own–operate–transfer scheme considering 18 years of private operation. The Mexican Federal Government provided 40% of the capital costs as a grant, whereas private financing provided the remaining 60%. Investment and operating costs are recovered by charging reclaimed water to the public power plant. The fee has three elements: one for the private return on investment and the other two for fixed and variable operating costs. The fee for the reclaimed water is 0.85 USD/m³, which is 33% cheaper than groundwater, which in fact, is no longer allowed to be extracted.

An additional benefit was the restoration of the ecosystem in the Tenorio Tank, which initially received untreated wastewater. The tank now works as an artificial wetland that polishes and improves water quality before its use for irrigation. Recently, migratory birds have returned to nest in its surroundings, which is an ecological breakthrough.

Source: with information from Rojas *et al.* in US-EPA (2012a).

population is aware of the benefits of reuse, but simply because wastewater from cities is available for farmers to irrigate. Farmers soon understand that wastewater is available year-round and it contains nutrients, representing savings on the use of fertilisers and the quality of the soil is improved thanks to its organic matter content.

Controlling the use of wastewater to irrigate agricultural fields is important. Thebo *et al.* (2017) estimated that the global area with crops irrigated with water influenced by wastewater (both treated and untreated) is 35.9 Mha, 82% of which are located in countries where less than 75% of wastewater is treated. This practice has been and is still growing globally (Qadir, 2022; Qadir *et al.*, 2020).

The WHO guidelines from 2006 (WHO, 2006) on the reuse of water in agricultural land[14] have greatly improved this practice in a safe and sustainable manner. These guidelines have adopted a comprehensive risk assessment and management framework, which identifies and distinguishes vulnerable communities and considers the trade-offs between potential risks and nutritional benefits in a broader development context. As such, the WHO approach recognises that conventional wastewater treatment may not always be feasible worldwide to achieve the desired quality levels, particularly in resource-limited settings, and offers alternative measures that may reduce the disease burden of wastewater use. These measures are known as risk mitigation measures as part of a multi-barrier approach (Box 3.5). To protect farmers, awareness campaigns about the invisible risk of pathogens should accompany the promotion of protective clothing (boots, gloves, etc.), hygiene, and, where possible, a switch to irrigation methods that minimise contamination and human exposure, such as drip irrigation.

The primary exposure or transmission of risks involves a direct contact with faeces, contaminated water and person-to-person contact. Airborne transmission due to the inhalation of very fine particles of wastewater that is applied through a sprinkler irrigation method is unlikely (WHO, 2006). Secondary transmission refers to transmissions from entities that have been exposed to polluted water, such as food when washed, and vectors.

----

[14] The specific approach used by the WHO (2006) guidelines is to (1) define a maximum tolerable additional burden of disease, (2) derive tolerable risks of disease and infection, (3) determine the required pathogen reduction(s) (s) to ensure that tolerable disease and infection risks are not exceeded, (4) determine how the required pathogen reductions can be achieved and (5) establish a system for verification monitoring.

Box 3.5 Example on the use of WHO guidelines to set a multiple barrier approach to control the use of wastewater for irrigation

In sub-Saharan Africa less than 1% of wastewater is treated; however, an important share of it is used to irrigate crops. Due to financial reasons, the traditional approach to implement only water quality standards to reduce risks cannot be applied. Monitoring of water quality is also difficult due to the nature of the agricultural practice: informal, small-scale and with many farms spread throughout and around the cities. For this reason, WHO (2006) guidelines were used as a viable alternative to manage health risks. This multi-barrier hazard analysis and critical control points approach is applied when treatment cannot meet water quality thresholds. Some risk reduction options, most of which have been tested in Ghana, are shown in Table B3.1. These options can easily be combined to achieve optimal contamination reduction. For example, farm-level water treatment can be combined with good irrigation techniques, better management in markets and household vegetable washing for further cumulative reduction in contamination.

Table B3.1 Non-conventional health protection control measures and associated pathogen reductions.

Control Measure	Pathogen Reduction (Log Units)	Notes
A. Wastewater treatment	6–7	Reduction of pathogens depends on type and degree of treatment selected.
B. On-farm options		
Crop restriction (i.e. no food crops eaten uncooked)	6–7	Depends on (a) effectiveness of local enforcement of crop restriction, and (b) comparative profit margin of the alternative crop(s).
On-farm water treatment		
(a) Three-tank system	1–2	One pond is being filled by the farmer, one is settling and the settled water from the third is being used for irrigation.
(b) Simple sedimentation	0.5–1	Sedimentation for ~18 hours
(c) Simple filtration	1–3	Value depends on filtration system used.

(Continued)

Table B3.1 Non-conventional health protection control measures and associated pathogen reductions *(Continued)*.

Control Measure	Pathogen Reduction (Log Units)	Notes
Method of wastewater application		
(a) Furrow irrigation	1–2	Crop density and yield may be reduced.
(b) Low-cost drip irrigation	2–4	Reduction of 2-log units for low-growing crops, and reduction of 4-log units for high-growing crops.
(c) Reduction of splashing	1–2	Farmers trained to reduce splashing when watering cans used (splashing adds contaminated soil particles onto crop surfaces which can be minimised).
Pathogen die-off (cessation)		
	0.5–2 per day	Die-off between last irrigation and harvest (value depends on climate, crop type, etc.).
C. Post-harvest options at local markets		
Overnight storage in baskets	0.5–1	Selling produce after overnight storage in baskets (rather than overnight storage in sacks or selling fresh produce without overnight storage).
Produce preparation prior to sale	1–2	(a) Washing salad crops, vegetables and fruits with clean water.
	2–3	(b) Washing salad crops, vegetables and fruits with running tap water.
	1–3	(c) Removing the outer leaves on cabbages, lettuce and so on.
D. In-kitchen produce-preparation options		
Produce disinfection	2–3	Washing salad crops, vegetables and fruits with an appropriate disinfectant solution and rinsing with clean water.
Produce peeling	2	Fruits, root crops
Produce cooking	5–7	Option depends on local diet and preference for cooked food.

Source: With information from Amoah *et al.* (2005) and US-EPA (2012a, 2012b).

3.3.4 Risks to sanitation workers

Sanitation decision makers need to be aware that their business presents risks to its workers. Besides the common risks due to accidents in industrial facilities, there are other risks related to the nature of their work. These risks have to do with the management of wastewater, sludges and faecal material, all containing microbes and other types of organisms that are sources of infection. Furthermore, there is the risk from biodegradable organic matter containing sulphates that are the origin of poisonous gases. These risks, although well-known and documented in literature, still cause several unnecessary work diseases and deaths in many parts of the world, in the Global North and Global South. It is important to train sanitation workers at all levels on all the risks and ways to control them. Clearly this will demand special efforts in rural and low-income areas, where the risks are considerably higher.

Sanitation work is an essential public service, but often sanitation workers are employed in the informal sector and are some of the most vulnerable workers (WHO, 2022). Approximately, two-thirds of countries have national laws or regulations in place to ensure the health and safety of: toilet cleaners; faecal sludge emptying workers, transport and treatment workers and sewage and WWTP workers (WHO, 2022). However, these laws and regulations are applicable to the formal sector and likely miss the informal workforce working outside any legal and regulatory protections. The Glass survey 2021/2022 reports that only 50% of countries have operational guidelines for worker's health and safety fully in place, 37% have mechanisms to check their compliance, 41% provide occupational health and safety training for workers, only 29% have sufficient equipment to enable safe operation and 35% sufficient personal protective equipment (WHO, 2022). Only in India, there is an estimated 1.2 million scavengers working in sanitation-associated tasks. Most of them have working conditions that have remained virtually unchanged for several decades. They are not only a socially rejected class but also exposed to several health risks. These health hazards include exposure to methane and hydrogen sulphide gases (Tiwari, 2008).

3.3.4.1 Infection risks

Sewer workers are more likely to develop certain infectious diseases than the general public. Exposure may occur by (Tiwari, 2008):

- Hand-to-mouth contact while eating, smoking, drinking or wiping the face.
- Breathing in mist, dust or aerosol.
- Skin contact through scratches or cuts.

The most common infectious diseases contracted by sanitation workers include:

- Hepatitis: Although preventable through vaccination, hepatitis is still one of the most common infectious diseases sewer workers develop. Research suggests that exposure to sewage is linked to a higher risk of contracting hepatitis B.

- Leptospirosis: This disease affects people who are in contact with animals and their waste. Urine from rodents and other animals in the area can contaminate the sewers, putting workers at risk of developing leptospirosis.
- *Helicobacter pylori*: This bacterium is now considered an important risk factor for gastric cancer and is considered a class I carcinogen by the International Agency for Research on Cancer.

3.3.4.2 Risks associated with gases

The gases that are a source of concern are methane, hydrogen disulphide, carbon monoxide and ammonia; all of which cause breathlessness, sore throat, chest tightness, cough, sweating, thirst, loss of libido and irritability. The greater the exposure, the more severe the symptoms (Kingsley, 2021; Watt *et al.*, 1997). Most of the gases mentioned, such as methane, are toxic and explosive and are well handled in WWTPs as part of the standard safety procedures.

However, this is not the case for hydrogen sulphide (H_2S), which is present not only in WWTPs, but also in the sewer network, pump stations, cesspools or any other closed environment where wastewater and sludge flow is slow or stagnant (Austigard *et al.*, 2018). It is really worrying that several sanitation workers still die in many countries due to the inhalation of hydrogen sulphide. H_2S is a gas that is produced through bacterial processes in which the organic material containing sulphur or the sulphates commonly found in wastewater and faecal sludge are decomposed. The gas is colourless, poisonous, corrosive and flammable. It smells like 'rotten eggs' and is slightly denser than air, thus it tends to accumulate at the bottom of poorly ventilated spaces. A mixture of H_2S and air can be explosive (Greenwood & Earnshaw, 1997; Knight & Presnell, 2005). H_2S is an important cause of work-related sudden death. The gas is particularly insidious due to the unpredictability of its presence and concentration, and its neurotoxicity at relatively low concentrations, causing olfactory nerve paralysis and loss of the warning odour (NRC, 2009). Inhalation of H_2S had resulted in about seven workplace deaths per year in the USA (2011–2017 data), second only to carbon monoxide (17 deaths per year) (Knight & Presnell, 2005). There are several examples in which several people die due to single case exposure to H_2S. The pattern is the same: when the first person enters an unventilated area with high content of this gas and does not come back, a second person goes to the rescue with no precautions and is also gets affected (Box 3.6).

3.3.4.3 The sanitation working social environment

Sanitation system workers are often stigmatised and marginalised, facing difficult conditions; health risks undignified and unacceptable environments in unsanitary and unregulated environments. Their work can expose them to hazards and social rejection, which can lead to drug and alcohol abuse to cope with the dehumanising conditions of the worst types of sanitation work in the world. Working conditions must be gradually formalised to protect health and safety and offer decent working conditions (UN-ESCAP, UN-HABITAT & AIT, 2015).

Box 3.6 Examples of reported deaths caused by H_2S.

- In 2014 workers at the Promenade shopping centre in North Scottsdale, AZ, USA died after climbing into a 4.5 m high chamber without wearing personal protective gear. 'Arriving crews recorded high levels of hydrogen cyanide and hydrogen sulphide coming out of the sewer'.
- In 2017, three utility workers in Key Largo, FL, died one by one within seconds after descending into a narrow space beneath a manhole cover to check a section of a paved street.
- In 2019, an employee of Aghorn Operating Inc., Odessa, TX and his wife were killed due to a water pump failure. The worker died while responding to an automated phone call he had received alerting him of a mechanical failure in the pump, while his wife died after driving to the facility to check on him.

Source: with information from Knight and Presnell (2005).

Analysing the neglected situation of sanitation workers, the World Bank (2019a), WHO and the International Labour Organization and WaterAid suggested four priority areas of action:

(1) Reform policy, legislation and regulation to acknowledge and professionalise the sanitation workforce along the sanitation service chain.
(2) Develop and adopt operational national and local guidelines to assess and mitigate the occupational risks for public and private sanitation service providers, provide training and ensure technology and personal protective equipment is available for all the workers along the sanitation chain.
(3) Advocate for sanitation workers and promote their empowerment to protect worker rights and amplify worker voices through unions and associations.
(4) Build the evidence base to address issues around quantifying the sanitation workforce and to document challenges workers face.

3.3.5 Sanitation and solid wastes

In literature this aspect is barely covered, possibly because this is not a problem concerning high-income countries. Any person who has worked in treatment plants of sewerage systems in regions where the service of solid waste collection is poor knows that at open and closed sewage channels and at the entrance of WWTPs a significant amount of a diverse variety and sizes of solid wastes are collected. These solid wastes overload the pretreatment steps (racks or degritters), and it is frequently necessary to install specific equipment or manual practices to deal with them.

Another issue, which may be less evident to practitioners, is that in countries with tropical storms (including several Global South countries), sewers have to be built with a high capacity to collect the abundant rainwater that falls in short periods of time. As a result, for an important part of the year, sewers convey a low flow and solid sedimentation is a frequent problem. To maintain their capacity, sewers must be cleaned before the rainy season, removing large volumes of sediments and wastes. For instance, in Mexico City, cleaning the entire sewerage system represents an amount of 2.8 Mm^3 sediments, 0.85 Mm^3 per year. The material that is extracted has characteristics similar to a primary sludge, and thus it must be treated prior to its disposal or use as landfill cover cells (Jiménez *et al.*, 2004).

3.3.6 Water rights

Water rights is a subject frequently overlooked by several policy and decision makers working in water and sanitation services. As already mentioned, several middle- and low-income countries are located or have significant parts of their territory in arid or semiarid areas, thus water resources are scarce. These countries, in contrast to countries where water is not a limiting resource, have strict water rights/concession systems to grant water to users. When the strategies to increase sanitation coverage are coupled with water reuse, it is important to keep in mind that in water-scarce areas it is highly probable that the non-treated wastewater is being used with or without permission by somebody (most probably a farmer). Therefore, decision of treating the water to reuse is not a straightforward decision and negotiations need to be undertaken with people who are already using water (Chapter 4, see Box 4.1, example 1). There are several examples around the world where farmers are using non-treated wastewater for irrigation (Jiménez *et al.*, 2010), and when the government starts sanitation programmes planning to provide them with water of better quality or to sell it to a new user (frequently an industrial user), farmers oppose the project (Jimenez, 2011). Even if farmers do not have the formal concession to use the water, if they have been using it for years, they have customary rights. Therefore, before undertaking sanitation programmes coupled to reuse, it is first necessary to check the current water use in the region and plan on awareness raising, communication and negotiation campaigns accordingly.

3.3.7 Climate change

The water sector has always been affected by changes in weather, therefore it has experience in managing climatic events; however, the measures in place may be surpassed by the effects of the climate change (Caretta *et al.*, 2022; Jiménez Cisneros *et al.*, 2014; Kundzewicz *et al.*, 2008). According to the Intergovernmental Panel on Climate Change (Caretta *et al.*, 2022; Jiménez Cisneros *et al.*, 2014), an increase in temperature, floods, droughts, extreme rain events and an increased water scarcity provoked by anthropogenic warming affect the sanitation service. Thus, it is necessary to consider such impacts when designing, operating and managing the sanitation service to minimise risks. Sanitation is considered as an option for adaptation[15]. For instance,

---

[15] 'Climate change adaptation' is defined as an activity that intends to reduce the vulnerability of resilience, through increased ability to adapt to, or absorb, climate change stresses, shocks and variability and/or by helping reduce exposure to them WHO (2022a).

it contributes to reduce the overall disease burden experienced by poor and marginalised communities, meaning they are better prepared to cope with the impacts of climate change or it can help when coupled with reuse to provide additional water sources in water-stress areas. Furthermore, it contributes to the mitigation of greenhouse gases (GHGs) in such a way that even funds to face climate change are available (Caretta *et al.*, 2022; Coninck *et al.*, 2018; Jiménez Cisneros *et al.*, 2014)[16].

3.3.7.1 *Effects on sanitation*

An increase in temperature affects the speed of biological and chemical reactions. An increase in water temperature in the context of climate change accelerates biodegradation which is positive for biological wastewater, sludge and faecal matter treatment processes but negative for the release of H_2S from the associated compounds. At a higher water temperature, the amount of oxygen that can be dissolved in water and is available for aerobic biological reactions is reduced and under extreme conditions this can be a limit for aerobic biodegradation. The resistance of some viruses to disinfectants has increased with temperature, whereas for some species of bacteria their persistence has decreased (Carratala *et al.*, 2020; Gundy *et al.*, 2009). Higher atmospheric temperature raises the demand for water for irrigation and when clean resources are not available the non-controlled reuse of wastewater might increase (Jiménez Cisneros *et al.*, 2014).

Hydrological extremes are projected to become more frequent and more intense due to warming conditions. Flood risks are expected to double when the temperature rises between 1.5 and 3°C (Dottori *et al.*, 2018) and an estimated 120–400% increase in population at risk is expected from river flooding when the temperature rises between 2 and 4°C, respectively (Caretta *et al.*, 2022). During floods and extreme rain events, the sanitation infrastructure often fails (due to damages by water or electricity shortages) and stops working. It takes time to recover adequate operating conditions, as damages range from a diverse degree of repairs up to rebuilding the entire infrastructure. It has been documented that drinking water and sanitation coverage decreases during floods (Caretta *et al.*, 2022).

Extreme rain events can cause an overload and overflow of sewers. Wastewater outflows are associated with an increased risk of gastrointestinal illness through the contamination of drinking water sources (Jagai *et al.*, 2015; Khan *et al.*, 2015). Floods intensify the mixing of floodwater with wastewater and the redistribution of pollutants (Andrade *et al.*, 2018). Sea-level rise affects the functioning of sewers, discharging wastewater to the sea or water bodies associated with it. Higher water scarcity caused by anthropogenic warming results in less water for sanitation facilities and a higher pollutant concentration in the wastewater to be treated.

---

[16] 'Climate change mitigation' is defined as an activity that contributes to the objective of stabilisation of greenhouse gas concentrations in the atmosphere at a level that would prevent dangerous anthropogenic interference with the climate system by promoting efforts to reduce or limit greenhouse gas emissions or to enhance greenhouse gas sequestration (WHO 2022a).

3.3.7.2 Adaptation

Similar to any other infrastructure, sanitation needs to be more resilient to climate change. It is important when selecting approaches or technologies for sanitation, especially in the context of climate change, choosing low-regret or hybrid solutions (Cutter *et al.*, 2012). Low-regret solutions are adaptation measures that besides being able to cope with the effects of climate change, serve to increase the coverage of the sanitation services or to improve its quality and therefore they are needed independently of warming impacts. A hybrid approach is recommended for urban and peri-urban areas. Hybrid solutions combine 'hard' engineering structures (grey) with managed or restored biophysical systems (green and blue) (Ngoran & Xue, 2015; Palmer *et al.*, 2015). One example is the use of sustainable urban drainage systems which reduce flooding, improve stormwater quality and reduce the urban heat island effect (Caretta *et al.*, 2022).

An additional aspect concerning sanitation and adaptation is the need to implement sanitation facilities (from toilets to the treatment and disposal of faeces/wastewater) in shelters hosting victims of fires or floods or displaced people due to internal and international climate change (Missirian & Schlenker, 2017; Rigaud *et al.*, 2018). This means being able to provide the service, and having an additional water source.

3.3.7.3 GHG emissions and mitigation

The sanitation services are directly and indirectly responsible for GHG emissions: directly through the breakdown of excreta and any other biodegradable material (Caretta *et al.*, 2022); indirectly because of the use of energy produced with fossil fuels. The latter may be reduced by using green energies such as solar power which works efficiently for small installations in rural areas.

Regarding direct emissions, the main concern is methane (CH_4), which is the second most important anthropogenic GHG after carbon dioxide (CO_2), and is responsible for over a third of the total anthropogenic climate forcing. Methane is also the second most abundant GHG, accounting for 14% of global GHG emissions. Methane is considered a 'short-lived climate pollutant' (12 years), and although it is emitted in smaller quantities than CO_2, it traps more than 25 times the amount of heat (Global Methane Initiative, 2013)[17]. Globally, methane from wastewater contributed 512 MMTCO$_2$[18] in 2010 (US-EPA, 2012a, 2012b).

Methane is emitted during the handling and treatment of municipal wastewater and faeces through the anaerobic decomposition of organic material.

---

[17] The Global Methane Initiative launched in 2004 is the only international effort to specifically target the abatement, recovery and use of the GHG methane by focusing on the five main methane emission sources: agriculture, coal mines, municipal solid waste, municipal wastewater and oil and gas systems.

[18] Million metric tonnes of CO_2 equivalent: metric tonnes of carbon dioxide equivalent or MTCO$_2$e is a unit of measurement. The unit 'CO_2e' represents an amount of a GHG whose atmospheric impact has been standardised to one unit mass of carbon dioxide (CO_2), based on the global warming potential (GWP) of the gas.

Most Global North countries rely on centralised aerobic wastewater treatment systems to collect and treat municipal wastewater. These systems produce small amounts of methane emissions, and also large amounts of biosolids are produced that can result in high rates of methane emissions when treated. In the Global South, the few collection and wastewater treatment systems that exist tend to be anaerobic, and thus result in greater methane emissions (Global Methane Initiative, 2013). These systems include lagoons, up-flow anaerobic, sludge blanket reactors, anaerobic reactors, septic tanks and latrines.

Besides reducing GHG emissions from the sanitation services, it is possible to capture and use the methane produced at wastewater treatment facilities (Box 3.7) as it is a source of energy that supports energy independence, replacing fossil-fuel use.

Box 3.7 Examples of WWTPs reclaiming methane for on-site power production and consumption

La Farfana Wastewater Treatment Plant (WWTP), Santiago, Chile. (GMI, 2013)

The La Farfana WWTP, managed by Aguas Andinas, treats more than 60% (8.8 m^3/s) of the wastewater in the Santiago Metropolitan Area. This project upgrades biogas from the anaerobic digesters to town gas quality. Town gas quality is achieved using a treatment train consisting of compression and dehydration to eliminate humidity, a bioreactor and a scrubber that removes 95% of the hydrogen sulphide (H_2S), and a thermal oxidiser that removes CO_2 and traces of oxygen and nitrogen in the gas. Afterwards, the treated gas is sold to the Metrogas Town Gas Plant located 13.8 km west of the Farfana WWTP. The project was registered as a Clean Development Mechanism project in 2011 and is expected to yield reductions of 26,000 metric tonnes of CO_2. This is an equivalent reduction per annum of what would be produced using fossil fuels.

Arrudas WWTP, Sabará, Brazil. (GMI, 2013)

The Arrudas WWTP serves around 1.5 million people from the Metropolitan Region of Belo Horizonte. The WWTP is a 3.3 m^3/s (4.5 m^3/s final design flow) activated sludge plant that utilises anaerobic digesters for sludge treatment. Since 2012, the biogas produced from the anaerobic digesters is treated to remove H_2S and used to generate heat and power for the WWTP in a combined heat and power (CHP) system. The CHP system consists of 12,200 kW microturbines, for a total electricity generating capacity of 2.4 megawatts. The electricity produced is used completely on-site and meets 90% of the requirements of the WWTP. Hot exhaust gases from the microturbines flow through heat exchangers to heat-recirculated sludge from the digesters to optimise biogas production.

Biogas and power generation and benefits in Brazil

Passos *et al.* (2020) calculated the mass balance of biogas production and thermal energy generation that could be achieved via the implementation of small-scale anaerobic-based systems to address the energy needs for around 8.3 million Brazilian inhabitants living in rural areas that are within urban areas, isolated rural areas with large settlements and isolated rural areas with small settlements, according to the Brazilian Institute of Geography and Statistics cited in the same study.

The thermal energy available in the biogas would be enough to sanitise all the sludge produced in the sewage treatment plants (STPs), making this biosolid material available to small-scale farmers or even to encourage the practice of family farming close to the treatment plants. Besides contributing to closing the nutrient (N and P) cycles and lowering the production costs of agricultural products, there would be a huge indirect benefit derived from the shift in destination of this material, nowadays simply transported and disposed of in landfills. Moreover, the surplus of thermal energy (after sludge sanitisation) would be sufficient to supply more than 200,000 families in the northern region with biogas for cooking (replacing liquefied petroleum gas (LPG)), and around 40,000 families in the southern region with biogas for water heating (replacing electricity). Besides the direct social gains derived from supplying biogas for domestic uses in the vicinity of the STPs, there would be tremendous indirect gains related to the avoidance of GHG emissions, especially when biogas is used to replace LPG. In this case, we estimated negative (avoided) GHG emissions equivalent to 6.1 Gt CO_2eq/year.

There are several approaches for wastewater methane mitigation and recovery (Table 3.2) and options for the use of recovered methane (Table 3.3).

3.4 ELEMENTS TO CONSIDER FOR SANITATION PROGRAMMES AND PROJECTS

Once policies are set, it is necessary to apply them, developing programmes and projects. In the following sections the main aspects to consider when developing both are described.

3.4.1 Well-trained human resources

Among the main problems faced when programmes and projects are put in place are the lack of adequately trained human resources, insufficient financing and the need to fulfil unsuitable standards[19]. Insufficient financing

---

[19] Training refers to a system of educating current and future employees within a company or water/wastewater services. It includes various tools, instructions, and activities designed to improve performance at work. It assists employees and the company, as it is a way to increase people's knowledge and upgrade skills specific for their job.

Table 3.2 Wastewater methane mitigation and recovery approaches.

Recovery Approach	Description
Installing anaerobic sludge digestion (new construction or retrofit of existing aerobic treatment systems)	Anaerobic digesters are used to process wastewater or sludge and produce biogas, which can be used on-site to offset the use of conventional fuel that would otherwise be used to produce electricity and thermal energy.
Installing biogas capture systems at existing open air anaerobic lagoons	Biogas capture systems for anaerobic lagoons are the simplest and easiest method of biogas implementation. Rather than investing in a new centralised aerobic treatment plant, covering an existing lagoon and capturing the biogas can be the most economically feasible means to reduce methane emissions.
Installing new centralised aerobic treatment facilities or covered lagoons	Installing new centralised aerobic treatment systems or new covered lagoons to treat wastewater in place of less-advanced decentralised treatment options (or no treatment at all) can greatly reduce current and future methane emissions associated with wastewater. This option is most viable in areas with expanding populations that have the infrastructure and energy available to support such systems.
Installing simple degaussing devices at the effluent discharge of anaerobic municipal reactors	In several Global South countries with warm climate (e.g. Brazil, India, Mexico) anaerobic reactors – which are fed directly with municipal wastewater (e.g. (up-flow anaerobic blanket bioreactors (UASBs)), anaerobic filters, fluidised or expanded bed, baffled reactors) are being increasingly used for small and medium scale municipal WWTPs. In these systems, around 30% of methane produced is lost as dissolved gas in the treated effluent. A closed column with enough turbulence right after the reactor can capture a significant amount of methane, which may be used to generate power or directed to a flare.
Optimising existing facilities/systems that are not being operated correctly and implementing proper O&M	Optimising existing facilities and wastewater systems that are not being operated correctly to mitigate methane emissions is a viable alternative to installing new facilities or wastewater treatment processes such as anaerobic digesters. Proper O&M also ensures that facilities continue to operate efficiently, with minimal methane emissions.

Source: From Global Methane Initiative (2013).

will be discussed in Chapter 5, whereas the unsuitability of standards was discussed in Section 2.3.3.3 of Chapter 2 and education and training will be discussed in more detail in Chapter 4. In this section, we aim to highlight the need to have proper operators' training programmes for the entire sanitation chain. Even countries that are on track to fulfil SDG target 6.2 do not have sufficient trained human resources on sanitation, something that does not occur for the provision of drinking water. The GLAAS 2021/2022 country survey reports that only 14% of countries have sufficient amounts of skilled workers (WHO, 2022). Sanitation training programmes for the entire chain

Table 3.3 Wastewater methane use options.

Use Option	Description
Digester gas for electric and heat generation with a CHP	Facilities can use recovered methane as fuel to generate electricity and heat in a CHP system using a variety of prime movers, such as reciprocating engines, microturbines and fuel cells. On-site power production can offset purchased electricity, and the thermal energy produced can be used to meet digester heat loads and for indoor heating.
Digester gas for electricity or heat only	Facilities can use recovered methane as fuel to generate electricity and heat in a CHP system using a variety of prime movers, such as reciprocating engines, microturbines and fuel cells. On-site power production can offset purchased electricity, and the thermal energy produced can be used to meet digester heat loads and for space heating.
Digester gas purification to pipeline quality	Facilities can market and sell properly treated and pressurised biogas to the local natural gas utility.
Direct sale of digester gas to industrial users or electric power producers	Facilities can treat, deliver and sell biogas to a local industrial user or power producer, where it can be converted to heat and/or power. Digester gas can be used as a vehicle fuel. Facilities can treat and compress biogas on-site to produce methane of a quality suitable for use as fleet vehicle fuel.

Source: From Global Methane Initiative (2013).

rarely exist, or they are not fit for the tasks operators need to fulfil, particularly in low-income regions. Basic sanitation, sewers systems, WWTPs, sludge and faecal matter management and reuse projects employ specialised equipment and involve specialised methodologies for which operators need specialised training in order to operate the systems efficiently. When training programmes exist, but are not accompanied by reasonable salaries for operators, these eventually move to other sectors with better salaries. And, when countries do have skilled workers, those workers do not want to live and work in rural areas (WHO, 2022).

Human resources needed for sanitation involve proper educational programmes for policy development and planning, monitoring and evaluation, production of regulation, design of facilities and the operation and maintenance (O&M) (WHO, 2022). Future water professionals would need to have operational skills and skills to carry out diagnostic and monitoring activities, interpret the implications of test results, including economics and risk management (UN Water, 2015b).

3.4.2 Equity
3.4.2.1 Gender
Insufficient attention has been given to improving equity in access to water and sanitation services (Abedin *et al.*, 2019; Eakin *et al.*, 2020). Women are the most affected vulnerable group, as they often belong to at least two of the

different categories of vulnerable groups. Mainstreaming gender equity in sanitation policy is crucial to achieving sanitation for all, which in turn will go a long way in advancing many other parts of the 2030 agenda, especially in education and work (African Development Bank, UN, GRID-Arendal, 2020; UN Water, 2015a, 2015b). Without proper sanitation, clean water and hygiene facilities at home, in workplaces and in schools, it is truly difficult for women and girls to lead safe, productive and healthy lives (Caretta *et al.*, 2022).

Women not only need sanitation services, but they also need sanitation services adapted to them, for instance during their menstrual periods for which privacy is a requirement (UNICEF, 2019)[20]. Women need privacy in a different way than men. Women have specific hygiene requirements during menstruation, pregnancy and childbirth (Ellis *et al.*, 2016; Saleem *et al.*, 2019). To avoid experiencing uncomfortable situations, in many cases women prefer not to satisfy their necessities, risking their health. Furthermore, women and girls are more vulnerable to abuse and attacks when walking to a toilet or to open air defaecation sites. Less than two-thirds of countries reported that women's participation is specifically mentioned in national laws and policies. Notwithstanding, in the GLASS 2021/2022 report progress has been made considering women and girls in national sanitation policy and plans (71% of countries), but the effort to track and report results is lower (47% of countries) and even less for the financial support (21% of countries) (WHO, 2022).

Women's water rights are hampered by societal patriarchal norms that prevent women from accessing water and participating in water management (Caretta & Borjeson, 2015; Djoudi *et al.*, 2016; Sultana, 2018; Yadav & Lal, 2018). Numerous studies substantiate a male bias in information access, employment opportunities, resource availability and decision making in water-related adaptation measures (Huynh & Resurreccion, 2014; Sinharoy & Caruso, 2019). To ensure women's needs are considered gender-sensitive approaches need to be developed involving them in the design, implementation and management of facilities and institutions. The women work force is low, with 18% of the total workers in water utilities and 10% of all WASH positions in government ministries and national institutions (World Bank, 2019a, 2019b).

It is estimated that women produce around two-thirds of the food in most Global South countries, and yet they often do not have adequate access to land, water, labour, capital, technologies and other inputs and services. Where women are part of the decision-making process, sanitation becomes more relevant as they better understand the importance of having a healthy family (Box 3.8). This is because as women stay at home for the longest and are aware of the family needs, it is essential that they participate in the design and selection of domestic sanitation facilities, which contributes to the appropriation and proper functioning of the systems. Reliance on women's self-help groups and

---

[20] This is the main reason why the WHO-UNICEFJMP (2016), considers that improved sanitation refers to a facility that is not shared with other families, which should be viewed with a local cultural perspective, as in many low-income regions it is common for members of an extended family to share facilities.

> **Box 3.8 Empowerment of women in rural areas leading water reuse projects in Brazil**
>
> In 2015, the non-profit social organisation Colectivo Cunhã carried out a project to treat and reuse greywater to irrigate family gardens in the arid region of Western Cariri. The project was carried out in association with a collective (PATAC), the government (SEMEAR) and international development agencies (AECID/IICA/FIDA).
>
> In the rural environment of the Western Cariri, the labour division based on gender is very unfavourable for women. To strengthen their political and economic autonomy, the collective created groups of women to undertake social and political actions. Three pilot greywater treatment plants were installed to irrigate family gardens in the municipalities of Congo, Prata and Monteiro. During the project, awareness raising activities were performed on the potential to reuse water for agricultural production. The design and implementation of the project was participatory, cooperative and with the leadership of women to contribute to the empowerment of the community and as farmers. Besides empowering women, the project was very successful for the management of water and the contribution to food security (IICA, 2017).

associations has proven successful to develop and implement water adaptive responses (Caretta *et al.*, 2022). Research in South Asia shows that women's participation in sanitation programmes has led to increased coverage, better facility maintenance, increased hygiene awareness and lower incidence of faecal–oral transmitted diseases in the community (Fewtrell & Bartram 2001). Involving women in sanitation projects makes it easier to implement projects that impact other areas, such as food security for the family. However, involving women in projects does not mean considering them as a homogenous set of people. Women's local gender roles are not immutable or generalisable (Carr & Thompson, 2014; Djoudi *et al.*, 2016; Gonda, 2016; Sultana, 2018).

3.4.2.2 *Indigenous and local knowledge*
Indigenous Peoples have valuable, long-standing knowledge of their lands and waters, and yet they are often marginalised from water governance. It is necessary for governments and civil societies to recognise, integrate and listen to their concepts, needs and visions, to enrich water governance and build trust between stakeholders (UN Water, 2015a, 2015b).

Even though traditional communities used to live using the concept of what today we call 'circular economy' (Caretta *et al.*, 2022), colonisation mostly destroyed this approach replacing it with poor sanitation conditions. Their sustainable approach was part of the way in which Indigenous People saw and managed natural resources, including water. Applying decolonising approaches to the field of sanitation and freshwater management (Arsenault *et al.*, 2019; Wilson, 2019) will not only recognise the importance of local knowledge but may also enrich current practices for sanitation. In this sense the practices coupling sanitation with reuse makes sense and can facilitate culturally inclusive decision making and collaborative planning processes at the local and national levels (Harmsworth *et al.*, 2016; Parsons *et al.*, 2017; Somerville, 2014). The main challenges to overcome when managing sanitation and water reuse projects with Indigenous communities are:

- The need to adapt current legal framework to their traditional rules and processes to develop and implement projects.
- To integrate their vision within the market-based models of water rights regimes which can impede this type of exercises.
- To integrate the principle of gender inclusivity as we understand it, but also as they are adapting this concept in their own cultures. Having women on board is key as they hold much of the local and the traditional knowledge (Fauconnier *et al.*, 2018; James, 2019).

3.4.2.3 *Covering the needs of all vulnerable groups*

Countries are most likely to have measures for sanitation addressing people living in poverty (Figure 3.5; WHO, 2022). Other groups to consider are those living in remote or hard to-reach areas, people with disabilities, internally displaced persons, migrants and refugees living in camps, elderly people, people living in slums and informal settlements and religious groups. A challenging situation is faced in regions dominated by drug dealers or mafias of any kind (such a human trafficking); in many occasions these areas cannot be reached by governments and cases have been reported where the mafia groups provide public services. The idea in this section is to invite the reader to reflect on their own local conditions. Different countries, regions and societies have different vulnerable groups. For instance, in Colombia the areas affected by the 'Guerrilla' constituted a vulnerable group in which the vulnerability was the loss of the concept of having a government (Box 3.9).

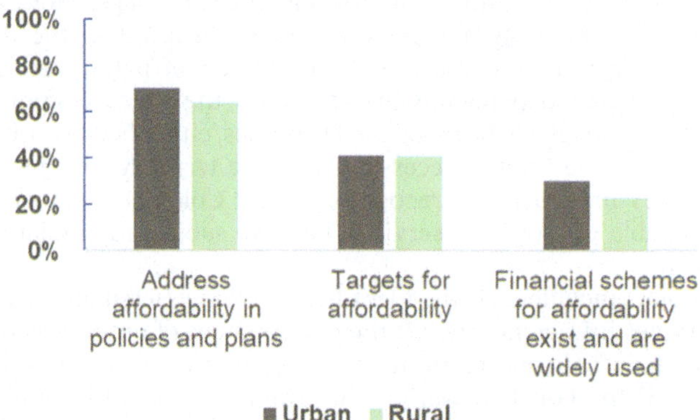

Figure 3.5 Percentage of countries having to address affordability for sanitation (*source*: WHO, 2022).

Box 3.9 Vulnerable group formed because of the 'Guerrilla' in Colombia needs water and sanitation services too

The United Nations Development Program (UNDP) is part of the international community participating in the process of peace agreement implementation, providing support for the development of strategies and actions to achieve greater social equity in the countryside and to contribute to the reincorporation of ex-combatants of the 'Guerrilla' into civilian life. An alliance between COLCIENCIAS currently the Ministry of Science, Technology and Innovation and UNDP was made in 2017 in the framework of the 'Science, Technology and Innovation Program in sustainable communities for peace'. This programme was designed to address the needs of the territories most affected by the 'Guerrilla' conflict and also to address the needs of communities with high environmental values, dispersed rural population and unsatisfied basic needs related to water, energy and productive processes.

COLCIENCIAS and the UNDP prepared a project for the 'Mariana Paez' Territorial Training and Reincorporation Area (ETCR) plus five villages from the municipality of Mesetas, covering an area of around 50,000 hectares and 468 families with cattle raising and coffee-farming activities. Drinking water and sanitation services were poor in the region because of the guerrilla occupation and the lack of government presence to provide public services. Many of the water and sanitation services that were built in the past were not operating due to lack of funds and technical knowledge to render the systems sustainable. As part of the Peace Programme, at each village a Community Action Board (JAC or Junta de Acción Comunal in Spanish) was formed with the purpose to improve the living conditions of inhabitants. The JAC members were also part of the Water Users Association (AUA, Asociación de usuarios de agua in Spanish) which was made responsible for the management of the drinking water and sanitation systems in each village. The JACs and the AUA are community associations created to assist in the decision-making participatory processes, take ownership of public services and even govern them independently in cases where the State does not provide or is not close enough to the associated activities, either because of the lack of resources or the limited accessibility to the territory. In this context, both the JACs and AUA were recognised by the Colombian government as entities capable of providing services, but also subject to regulations and supervision.

The project included: (1) the identification and assessment of available water services infrastructure; (2) the development of options to optimise the existing infrastructure or to propose new ones when it was not available; (3) the construction or improvement of facilities and (4) the involvement and training of the community to manage, operate and provide maintenance to the systems.

A combination of mostly decentralised and some centralised systems were selected for the region through a participatory process in which the community was closely involved. When centralised systems were used the JACs, through the AUA, were responsible for the management, hiring a person to provide O&M (a plumber), and to implement the financing mechanisms in which a monthly fee was charged to the users. An important component of the project was education and training. Local people were trained for the construction, management and maintenance of the sanitation systems. Awareness raising and education campaigns were performed for the community to take ownership of the project. At the end of the project the AUA, with the support of the community, was able to manage and maintain the systems autonomously without depending on the government.

Source: with information from Pontificia Universidad Javeriana-Universidad de La Salle (2021).

To succeed in addressing the needs of vulnerable groups, policy and decision makers must prioritise actions through measures that target these groups. They have to set measures in policies and plans, monitor progress to extend sanitation and ensure financial support. Additionally, reaching all does not only mean building facilities for the entire sanitation chain, but also to ensure that the services remain affordable. Affordability is also an obligation articulated within the UN recognition of the HRtWS and is part of the 2030 Agenda. Although affordability[21] itself is defined, addressed and monitored differently in each country, it is universally accepted as a key measure in leaving no one behind. Less than half of all countries have established targets for affordability of sanitation and only about a third have widely used financial schemes to support affordable sanitation services, with differences between urban and rural regions (WHO, 2022).

3.4.3 Use of innovative technologies
3.4.3.1 *Use of green and hybrid solutions*
Environmentally friendly systems or green infrastructure can facilitate a change towards a holistic response to manage the complete cycle of the water. To this end, green infrastructure must be evaluated from a technical, environmental, social and economic perspective. Natural infrastructure (green and blue) uses natural or semi-natural systems, for example, wetlands, healthy freshwater ecosystems and so on, to supply clean water, regulate flooding, manage sewage water, enhance water quality and control erosion. Compared with grey physical infrastructure, natural infrastructure is often more flexible, cost-effective and can provide multiple societal and environmental benefits simultaneously (Caretta *et al.*, 2022) (Box 3.10).

---

[21] In 2021, WHO and UNICEF reported recommendations on the ways in which affordability of WASH services can be monitored although it also recognises that more discussion on the subject at the international level is needed (WHO 2022).

> ## Box 3.10 Use of nature-based solutions in the management of water
>
> At the landscape scale, there is evidence that impacts from fluvial and coastal floods can be mitigated through nature-based solutions (NBSs) such as detention/retention basins, river restoration and wetlands. They are effective for floodplain restoration, natural flood management, managing sewage, regenerating biodiversity and 'making room for the river' measures.
>
> At the urban and peri-urban scales, NBSs increase resilience in cities by managing urban stormwater and contributing to mitigating the heat island effect. These also purify stormwater and provide additional hydraulic retention time that contributes to reducing urban floods. This is useful particularly to mitigate the effects of extreme precipitation events in urban areas. These types of solutions have been applied in cities such as New York and Copenhagen and have been built using different financing sources (including support from insurance companies; SWISS Re, 2024). Moreover, the criteria to select NBSs is increasingly based on integrated economic valuations that incorporate co-design with stakeholders. Although they have additional and different advantages, their performance to mitigate floods is limited when compared to grey infrastructure systems, especially when the volumes of stormwater to be managed is high. As mentioned before, hybrid solutions, combining green solutions with grey solutions, can provide the advantages of the two approaches, while giving additional time for more green solutions to be developed.
>
> *Source*: with information from Caretta *et al*. (2022), IUCN (2020).

3.4.3.2 Technology development

There is always room for new technologies, but for them to be mature enough to use they need to be developed through a process that might be long and comprise full-scale testing. If policy and decision makers would like to have new and better technologies, there is a need for them to support their development.

Adoption of technologies depends on the availability of finance and its appropriateness to the local context among other aspects. This means two things, that if poorly selected it will imply losing funds (and time) and that it must be proven effective under the same conditions where it is to be applied. The challenge is that to solve many public problems it is necessary to use new technology, but first this needs to be proven effective in the field and under similar conditions, regardless of its size, brand, country of origin or Curriculum Vitae of the commercialising company.

One approach to having more clarity on the use of new technologies is to be supported by academia and technology development programmes (Box 3.11), and when it is appropriate, to have the support from private companies to commercialise the inventions that have proven effective in the field (Box 3.12).

Box 3.11 Research and technology development: the need for a trigger to have direction

In 2011, the Bill & Melinda Gates Foundation established the *Reinvent the Toilet* challenge. The effort included working with researchers, scientists and manufacturers to develop safe sanitation solutions that work without relying on sewage systems or running water. In the decade of the launch of the challenge, the world has responded with the power of innovation. Scientists and engineers from across the globe developed hundreds of exciting ideas for how to design toilets that safely process human waste with little or no need for water or electricity. They created toilets that convert faeces into valuable resources, including fertiliser, clean water, electricity and other products. The next phase of the project is taking the best of these ideas to develop a low-cost reinvented toilet.

Source: with information from https://www.gatesfoundation.org/our-work/ programs/global-growth-and-opportunity/water-sanitation-and-hygiene/ reinvent-the-toilet-challenge-and-expo

Box 3.12 Thailand's reinvented toilet: a case study in innovation and commercial success

The Asian Institute of Technology (AIT) created four groundbreaking products. A hydro-cyclone cube, a system that utilises gravity and cyclone principles to separate solid waste from liquid, sterilised via heating, generating a reusable, pathogen-free by-product. A sanitiser truck, an adaptable vehicle equipped with a solid–liquid separator and a disinfection system, designed to reduce transportation and treatment expenses. A 'cess to fit' system, a retrofit tool capable of being integrated into existing cesspools to process faecal matter before its environmental discharge. A solar septic system uses a solar heating system to enhance pathogen elimination and organic matter biodegradation while improving the quality of septic tank effluent.

One design, the hydro-cyclone cube, renamed Zyclonic, was successfully commercialised through a partnership with SCG Chemicals Plc., credited as Thailand's first integrated waste treatment toilet proficient in waste separation and pathogen prevention. This Zyklon is a decentralised sewerage technology with self-contained toilets that help contain and treat pathogens without the need for water, sewer connection or electricity, and thus are suitable and sustainable for poor, urban settings. The successful introduction was signified by the establishment of pilot units in Rama IX's Khlong Phlabphla Community. Once the technology was proven in the field an intersectoral collaboration between the SCG Chemicals Co., Ltd. (private company), the Bill & Melinda Gates Foundation (international funder) and the AIT (academic) was initiated.

The progress and subsequent commercialisation of the reinvented toilet in Thailand highlights two crucial lessons. First, it demonstrated how innovation, collaboration, and sustainability could effectively address significant public health issues such as sanitation and yield commercial success. Second, it emphasised the need for community engagement and acceptance for these initiatives to thrive. The Zyclonic toilet case not only satisfied an immediate need but also facilitated community transformation, encouraged self-reliance and inspired individuals to aspire for improved community living standards.

Source: with information from ADB (2021a, 2021b).

Examples in Boxes 3.11 and 3.12 show that for technology development and to apply them support might come from different sources, but in most cases to be successful enterprises should be involved in the last step comprising commercialisation.

doi: 10.2166/9781789064049_00083

Chapter 4

The need to manage perception, attitudes and knowledge on sanitation

KEY MESSAGES:

- Sanitation and water reuse are not easy topics to understand and assimilate by users, politicians or donors. As a result, policy and decision makers need to make a special effort to render these topics accessible to all.
- Improving communication, better informing society and increasing education on sanitation and water reuse will assist in improving perception, attitudes and knowledge on associated projects improving the quality of participatory processes.
- Where the progress on sanitation services is low or poor and/or water reuse projects have caused damages and dissatisfactions, it is important to change the social perception encouraging a new vision that allows the execution of efficient and sustainable projects.
- There is a need for a paradigm shift for water reuse not to be perceived as a problem (social, health, environmental, economic, etc.) but as a resource of water and other products that when used appropriately produce well-being in the community.

4.1 THE NEED FOR SOCIAL PARTICIPATION

The 2030 Agenda for Sustainable Development (A/RES/70/1) makes it clear that without the participation of local communities the sustainable development goals (SDGs) – including SDG 6 'Clean water and sanitation' – will not be met. However, public participation processes for its implementation are still limited to some regions, countries, organisations and areas of expertise. This is due partly to the absence, in many regions of the world, of adequate legal, regulatory and institutional frameworks, and due to the social and cultural contexts. In addition, governments must be willing to promote public participation.

Participatory processes enable the community to be part of the planning, implementation and management of programmes. This participation takes place through different mechanisms including forums or community groups, through access to information, formal feedback to the provision of services, formal representation of users or communities in government processes for joint decision making and the involvement in the solutions of conflicts between users and service providers or between users. For the participatory processes to be useful it is essential that all actors are aware of their roles, their required and potential contributions, their challenges and their willingness to undertake changes, reforms and internal measures that will allow them to obtain the optimal conditions for sanitation services. During these processes, users and service providers need to consider their views on sanitation, identify what they need to learn and the aspects they need to understand from other people's perspective. This means that policy and decision makers need to know how to manage perception, attitudes and knowledge[1] on sanitation and water reuse to align objectives with the community.

Most countries (over 90%) reported in the GLASS survey to having procedures for public participation on water issues defined in law or policy. However, less than one-third actually put them in practice. Participation of users and communities is constrained by a lack of financial and human resources. Only 17% of countries indicated that they have over 75% of the financial resources needed to support participation. The lower the income level, the less likely that countries have sufficient financial resources in place (WHO, 2022).

4.2 PUBLIC PERCEPTION

The level of interest and the ways in which stakeholders participate in the planning or development of a project depend on their perception of it. Even when projects are technically well-planned and address sanitation societal needs, they can fail if public perception is not adequately considered (WHO, 2006). Public perception is the idea people have on a subject which does not necessarily correspond to the facts.

The perception that individuals, groups or communities have comes from the information available and personal experiences of the phenomena, its causes and effects (Bagheri *et al.*, 2008; Cookey *et al.*, 2016, 2020; Prinz, 1990). Perception is different from one person to another, although there may be common views on a subject for groups with common characteristics. Public perception depends on social, health, economic and environmental contexts, and is assessed through different ways, one of which is surveys (Bargh & Barndollar, 1996; Bargh & Ferguson, 2000; Dijksterhuis & van

---

[1] Some authors refer to this as KAPs, which stands for knowledge, attitudes and perception. The order of these concepts was altered in this book to follow the logic of process management.

Knippenberg, 1998). Keeping in mind that public perception is closely related to the local context, there are studies on sanitation and water reuse that assist in understanding the subject better (Kihila *et al.*, 2014; Kilobe *et al.*, 2013; Liberath Msaki, *et al.*, 2022; Mayilla *et al.*, 2017).

Understanding the perception of different social groups is useful to prepare meetings, communication and awareness raising and educational programmes. It also helps for the public to compare their beliefs with facts, understand opinions from different population groups and find ways to reach agreements. Understanding social perception on sanitation and reuse helps policy and decision makers identify preconceived notions and the sources of misconstrued visions of programmes and projects from: (a) users, (b) organisations concerned with the impacts of projects on the people, the culture, the environment and politics, (c) those that are not users but are impacted by the project (for instance, the people living next to the facilities or displaced by projects) and (d) the potential donors or politicians that are meant to support sanitation efforts (Box 4.1).

Box 4.1 Two case studies showcasing the relevance of understanding the public perception on sanitation projects

Example 1: Users' perception can be well grounded but not understood by decision makers due to their historical and educational background (Jiménez Cisneros, 1995; Jiménez & Chávez, 1998).

In 1992, the metropolitan area of Mexico City with 23 million inhabitants treated only 5% of its wastewater. Since the Spanish conquest (15th century) all the city's wastewater has been sent to a valley located north to the city. In the 19th century large sewers were built to transport this wastewater. The valley receiving the untreated wastewater is called the Mezquital Valley. It is in a semiarid region with only 550 mm precipitation per year and an evapotranspiration rate three times higher. The Mezquital Valley was a very poor region with low agricultural productivity because the soil was not adequate for agriculture and water for irrigation was scarce. The valley produced only one crop per year during the rainy season. Therefore, Mexico City's wastewater was soon used by local farmers for irrigation, increasing local crop productivity by 90 to 150% and harvesting three to four crops per year. The price of the land in the area receiving wastewater rose between three and six times. However, worm diseases among 4–16-year-old children increased by 16 times when compared to similar areas that used clean water for irrigation. Given the situation, the federal government started a project to treat and reuse Mexico City's wastewater. Farmers were opposed to the project, arguing they wanted to continue receiving the same amount of water and the 'substance' contained in it. This led to in-depth research to understand what 'the substance' was and to implement communication programmes reassuring farmers on the availability of water. The government had few possibilities of undertaking the project without farmer's approval, even if it benefited them. The Health, Environmental and Water Laws forbid the discharge of polluted water and its use for irrigation. In addition, the farmers had both the official concession of most of Mexico City's wastewater (they obtained it from the president in the 1960s) and the customary rights on it.

Research showed that due to the aridity of the region, farmers used a large amount of water for irrigation (2.5 m of water per year), which decreased soil salinity, and that the 'substance' they wanted to retain was the organic matter that improved the quality of the soil together with the nitrogen and phosphorus that fertilised it. This led to a change in the Mexican standards to include a quality of treated water aligned to agricultural irrigation needs, that is, limited removal of biodegradable organic matter and limited or no nitrogen and phosphorus removal, but the full elimination of helminth eggs which were causing worm diseases (helminthiasis).

Example 2: Social perception on sanitation is related to the technology and to the quality of services people receive (*source*: with information from Cookey *et al.*, 2016 and 2020).

Even though studies suggest that it is best to desludge septic tanks and latrines every 2–5 years, depending on the number of users, they are often not emptied at all or emptied after 20 years, when the content has solidified and becomes difficult to remove. When desludging is irregular or delayed it affects the effective functioning of the sanitation system leading to the overflow and spilling of its contents to the surrounding urban spaces, rural soil and water bodies, with negative impacts on health, the environment and the dignity of life. A perceptive study carried out to understand the reason why people from Thailand were reluctant to empty their on-site sanitation (OSS) systems found that these were technical and social. The technical barriers were related to poor construction, use of sub-standard materials and limited spaces to properly manage the latrines' content. These barriers are relatively easy to address by setting construction and operation standards, and ensuring they are enforced. The social causes were a lack of skilled operators and the users' lack of will to empty the latrines. For the first, educational and training programmes can be set-up, whereas the second – the active involvement of users – demands a better understanding of the reasons behind their lack of willingness to develop behavioural change programmes.

4.2.1 What we know about the perception on sanitation and water reuse

Although necessary, public perception studies are difficult to undertake, time consuming and expensive. Under these constraints, consulting literature is useful even if studies are not always applicable because they are for specific areas or come from general international surveys.

4.2.1.1 Per topic
4.2.1.1.1 Perception on sanitation

Worldwide, nowadays there is more awareness on health and environmental risks associated with the lack of sanitation, to the point that when the associated risks are perceived there is an impact on house prices, land and tourism (Xue *et al.*, 2022). This has been the reason for several international and national initiatives to improve water quality promoting sanitation and water reuse projects. However, the public perceives wastewater as a government problem rather than their own. Furthermore, even if people understand the need to build wastewater management systems, they do not necessarily agree on having the facilities close to the places where they live, on the type of processes selected, on the need to pay for the entire or part of the sanitation services and on being actively involved in the process. In fact, the public knows little about the sanitation chain and water reuse projects (Liberath Msaki *et al.*, 2022).

For basic sanitation, the situation is much more worrying. For many decades and in many sites around the world, non-sewered sanitation systems and OSS systems are not appreciated and people believe that sanitation projects are not good. There are many reasons for this. In some cases, sanitation systems are installed with little regards for the quality of the materials used, without following construction standards, are not properly operated hosting a vast diversity of insects and animals and being a source of bad odours, thus they are rejected by users (Box 4.2). This has given a bad reputation to certain types of technologies even 'when ecofriendly and adapted to developing countries' (Liberath Msaki *et al.*, 2022). This situation may happen because of the lack of funds and the aim to serve a wider range of people. However, in some cases it is due to the perception that poor technology and low-cost construction are for the poor. In many cases, the social and political conditions create situations in which part of the society receives services of good quality demanding no personal efforts for their operation whereas others have poor services and in addition to that they have to operate the systems themselves. Under these conditions sanitation increases societal divides.

Box 4.2 Importance of perception to advance on sanitation programmes in India

India has observed a substantial reduction in the proportion of the population practicing open defaecation; however, the practice remains high for a country with its economic development. The 2019 WHO/UNICEF Joint Monitoring Programme for Water Supply, Sanitation and Hygiene report estimated that 26% of India's population practiced open defaecation, 2% used an unimproved facility, 13% used a limited (shared) sanitation facility and 60% used at least a basic sanitation facility.

Since the 1980s, the Government has undertaken national sanitation campaigns to increase access to household latrines, primarily through financial subsidies. However, household access to a latrine does not guarantee its use; a significant proportion of latrine-owning households in India report members continue to practice open defaecation notably in rural areas (only 37% of rural households reported using an improved sanitation facility). Comparatively, neighbouring countries – Bangladesh and Pakistan –, despite having lower gross domestic products, have lower open defaecation rates (<1 and 10%, respectively). Pakistan and Bangladesh have the same or lower population percentages that use at least basic sanitation services compared to India (60 and 48% respectively) but have higher percentages using either an unimproved or limited sanitation facility.

Studies have found that despite the numerous efforts and investments made by the Government of India to increase latrine coverage, only modest reductions in open defaecation have been achieved with no measurable impact on child health. Barriers to latrine use include improper or incomplete latrine construction, lack of a nearby water source for post-defaecation cleansing, cultural beliefs around purity and pollution, fear of rapid pit filling and strong preference for open defaecation.

Source: with information from De Shay *et al.* (2020), Sclar *et al.* (2018, 2022).

The main personal reasons for rejecting sanitation solutions are discomfort, poor quality of facilities or lack of privacy. At the collective level, the predominance is for reasons affecting the environment due to the lack of support processes to dispose of faecal material or polluted water. One should ask what we would do if we were under the same conditions, in order to understand why many sanitation programmes have failed. People are rarely comfortable using time consuming, poorly built and ugly sanitation solutions, notably when considering that the time they need to spend for its operation and management could be used to continue working increasing the family income, spending time with family or, simply, having spare time for fun or relaxing. Services of good quality need to be sustainable; which also means the need to reduce equity gaps not only for sanitation but also for many other aspects. Figure 4.1 shows examples of the reasons for dissatisfaction and rejection of solutions that have been implemented to provide sanitation and perform water reuse. It can be observed that in general people are not against improving sanitation conditions or performing water reuse.

4.2.1.1.2 Perception on water reuse

Perceptions and public acceptance of water reuse are the principal factors for the successful implementation of this type of projects, regardless of the strength of scientific evidence in their favour (Liberath Msaki *et al.*, 2022; Michetti *et al.*, 2019). Although, nowadays, more people consider water reuse to be a vital endeavour towards protecting the environment, bringing economic gains and boosting agricultural productivity (Akpan *et al.*, 2020; Liberath Msaki *et al.*, 2022), the willingness to use reclaimed water depends on several underlying

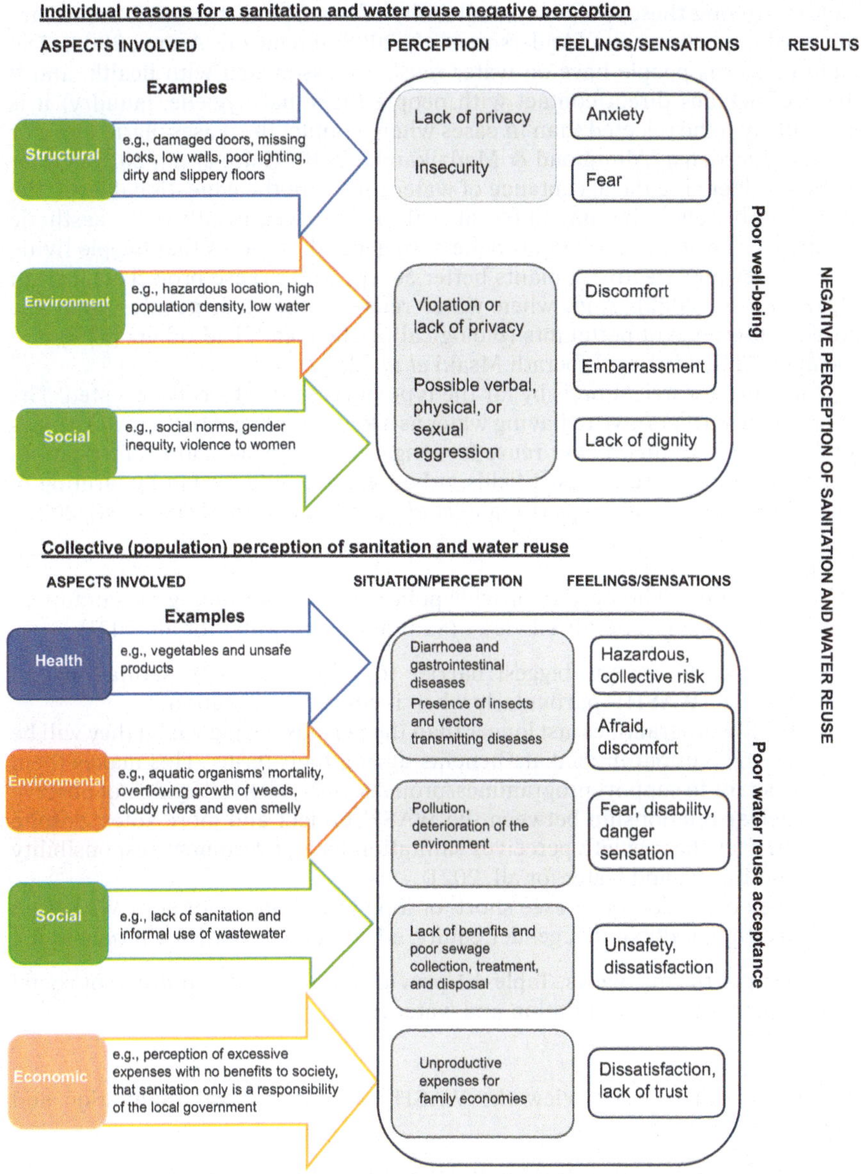

Figure 4.1 Reasons for the personal and collective 'negative perception' of sanitation and wastewater use.

factors. Among these, the local availability of water and the type of reuse are the most important ones (Anderson *et al.*, 2008; Jiménez & Asano, 2008). The main concerns people have on water reuse are associated with health, and if the use includes direct contact with people (personal hygiene, laundry) it is more likely to be rejected than in cases where contact decreases (agriculture or watering gardens) (Alhumoud & Madzikanda, 2010; Gatto *et al.*, 2015). Other factors influencing the acceptance of water reuse are the educational level, the costs and benefits, the magnitude of real or perceived health risks, aesthetic attributes of water and religious beliefs. In general, it seems that people living near wastewater treatment plants better accept water reuse projects (Liberath Msaki, *et al.*, 2022). And, when water reuse is objected, people signal the possible presence of pollutants (biological or chemical), bad odour (64%) and social or ethical issues (Liberath Msaki *et al.*, 2022).

The reuse for irrigation is by far the type of reuse most easily accepted. The importance farmers give to having water as a key component of their work plays a relevant role in this. Water reuse for irrigation covers agricultural irrigation, and irrigation of forests, sport fields, urban gardens and for family farming of vegetables and animal crops (Keraita *et al.*, 2015; Liberath Msaki, *et al.*, 2022).

4.2.1.2 Per social group

Politicians, those who need to provide political will and funding for sanitation, have the following generalised views (Sanitation and water for all, 2021):

- High costs are the biggest barrier to prioritise water, sanitation and hygiene (WASH)[2] particularly when it comes to sanitation.
- WASH programmes last longer than the periods during which they will be acting, without immediate benefits during their terms. This makes them hesitant to support programmes/projects. Specially, if they do not observe a clear relationship between the WASH project and more votes, despite the fact that society perceives sanitation as a government responsibility (Sanitation and water for all, 2021).
- They often do not foresee short- or long-term impacts of poor WASH on health, quality of life, gender equity, education or economic productivity.

To see politicians' views, Table 4.1 provides information on different social groups' perception on sanitation and water reuse.

4.2.1.3 Per sector

In general, other sectors view the WASH sector as follows (Sanitation and water for all, 2021):

- The WASH sector is insular and operates within a 'silo'.
- WASH is a complex and technical subject.

---

[2] Despite the fact that water experts with a technical background see a clear difference among the different elements of WASH, for most people there is not. More relevant is that in practice, sanitation cannot be fully dissociated from water supply and hygienic practices.

Table 4.1 Perception of sanitation and/or water reuse for different social groups.

Groups/Group Characteristic	Comment
Society at large and politicians	For most people, including politicians, sanitation is a very technical subject that in addition is not pleasant to talk about, especially in public spaces.
Policy makers and chief executive officers	Public perception on sanitation differs according to the sectors policy makers belong to, particularly for the health and water sectors. Health policy makers, although in favour of sanitation projects, are more reluctant or even clearly reject reuse projects as they perceive that if there is any negative impact on health, they will be the responsible for it. City majors and industrial managers in general are more in favour of water reuse. City majors directly receive the pressure from people to have additional water. Industrial managers consider that reclamation of water in their own facilities is their responsibility, as for water reuse (reusing water from another users) they are open to it, provided they can afford the additional treatment the water may need to be fit for their purposes.
Women and men	Females are more prone to reuse water for urban farming but more reluctant for domestic applications such as washing and cleaning the home. Women are more cautious and constant than men when there is recommended preventive measures (use of gloves and boots, for instance).
General public vs scientists	Because the topics are highly technical, the public tends to perceive risks differently from scientists. Among scientists, the perception varies according to their field of expertise. For example, ecologists tend to be concerned with the possible effects on humans and the environment while engineers tend to trust reliability of wastewater treatment. Agronomic specialists tend to better understand the advantages of reusing water.
Farmers	Farmers readily support water reuse for irrigation, minimising risks when water for irrigation is scarce.
Culture	In most of Africa the use of sewage or faecal sludge for agricultural purposes is discouraged, whereas in other regions, such as Asia, the practice is well recognised as economic and ecological.
Regions within a country	The perception on reuse is not the same throughout any given country. For instance, within the USA, states such as Florida, Arizona or California, where water is scarce, promote reuse projects, legislation and policies whereas others located in the northeastern side of the USA, where water is abundant, do not.
Age	The youth and people of working age (36–53 years) are more positive and knowledgeable about water recycling and reuse than older people (more than 65 years old).
Degree of education	Higher education degrees help people to be more positive on sanitation and water reuse projects.
Ethnicity	Some ethnic groups mistrust the government much more than the general population, leading to a lower acceptance of water reuse as having confidence in the government running projects is key.
Religious	For Islamic communities, water reuse is acceptable if, for example, the wastewater undergoes a purification or dilution procedure before reuse.

Source: partly, with information from Robinson *et al.* (2005); Jiménez and Asano (2008); Keraita *et al.* (2010); US-EPA (2012a, 2012b); Cotruvo *et al.* (2013); Gu *et al.* (2015) and Fielding *et al.* (2018).

- WASH contributes differently to their own objectives and priorities.
- The three components (water supply, sanitation and hygiene) have a different value/importance.
- Only the human rights, the humanitarian response and the health sectors value sanitation highly.
- The education, human rights and humanitarian response sectors see WASH as part of human security and human rights.

The perception of each sector as part of sanitation is listed below (Sanitation and water for all, 2021), whereas their perception on sanitation is presented in Table 4.2:

(a) *Economic development sector*, For this sector, hygiene is less important than water and sanitation, and they believe it is important to have robust data showing the fiscal benefits to implement sanitation and to explain the risks of inaction, in particularly the financial costs.

Table 4.2 Top three sectors priorities and perception of sanitation benefits.

Sector	Top Three Sector Priorities	Perception of Wash Benefits
Economic development	(1) Sustainable economic development (2) Financial inclusion (3) Access to education and skills development	(1) Improve health of workforce (2) Improve workplace conditions (3) Improve gender equity and foster a more inclusive workforce
Health	(1) Disease prevention (2) Access to WASH (3) Equitable access to healthcare	(1) Preventing disease (2) Protecting marginalised groups (3) Enhancing nutrition
Education	(1) Quality of education (2) Access to education for all (3) Gender equity	(1) Improve gender equity in schools (2) Improved educational attainment (3) Improve educational attendance
Human rights	(1) Gender equity (2) Children's rights (3) Inclusive sustainable development	(1) Elevate the rights of marginalised groups (2) Ensuring the right to healthy life/SDG 3 (3) Build inclusive communities
Humanitarian response	(1) Protecting human rights (2) Public health in crisis zones (3) Climate change resilience	(1) Disease prevention (2) Protection of vulnerable groups in crisis scenarios (3) Stronger community resilience
Climate change	(1) Resilience to climate change (2) Ecosystem conservation (3) Climate change mitigation	(1) Improving the health of communities and ecosystems (2) Improving food security (3) Helping vulnerable communities adapt to climate risks

Source: with information from Sanitation and water for all (2021).

(b) *Health sector*: Their concerns are the short-term risks, which can be interpreted as (microbial) diseases. WASH is among their highest priorities due to its link with health, perceiving its improvement as a means to bring benefits to their sector. However, they observe hygiene and sanitation as slightly more important than water and tend to have similar views among them on the ways in which sanitation practices should be improved.

(c) *Education sector*: Hygiene is considered a highly important element of their work.

(d) *Human rights sector*: Sanitation is relevant because of its contribution to creating safer societies, particularly for marginalised groups, and for the safety of women. They consider WASH a way to contribute to the SDGs on health and well-being and are concerned by the long-term effects of poor WASH on human security.

(e) *Humanitarian response sector*: The importance is given to the short-term impacts of WASH on the people and on safety. Clearly, they perceive the link of WASH with human rights, health and climate. They believe that there is a need for more awareness raising on hygiene and sanitation (compared to water supply), and to raise political support on the WASH agenda.

(f) *Climate change sector*: Perceives that WASH can contribute to priorities for adaptation; however they do not consider there are direct links between WASH priorities and theirs, other than contributing to reduce people's vulnerability. Among the components of WASH, they think water *as a resource* is the most important one.

4.2.1.4 Per region

Africa makes a strong link between improved sanitation and disease prevention, compared to other regions, considering that all of its components contribute to better education attendance. In general, the public does not feel the presence of the government is significantly necessary to address the issue. In Asia, sanitation is considered as a key component for community well-being, protection and safety, as opposed to its financial impact. The connection between WASH and climate resilience is established, although the impact of sanitation and having access to clean water is unclear. In Latin America the core aspects of WASH are related to water, focusing on its impacts on the community and human benefits rather than on financial aspects. WASH is perceived as a way to protect and support vulnerable groups and considers its benefits on climate change to be low. Finally, the European and North American regions are more concerned on the impact of WASH on financial aspects demanding hard data to support its implementation. Sanitation is perceived in a broader context as part of human rights, and the benefits it brings to tackle gender inequities in education are given high importance (Sanitation and water for all, 2021).

4.2.1.5 Recommendations to understand stakeholders' perception

Policy and decision makers cannot know in advance the societal, politicians and donors' perception on sanitation in their local context (WHO, 2006). Thus, it is recommended to look for the precise information, keeping in mind that the subject is highly technical and when enquiring, it is always important to define terms (Liberath Msaki *et al.*, 2022). To understand the perception of stakeholders, professional studies can be undertaken, but this is not always economically feasible. Thus, there is a need to find alternative ways to get information through meetings, research centres or the media, for instance, on what stakeholders think about:

- The need for an adequate collection and management of wastewater to avoid the pollution load causing health or environmental problems.
- The willingness to support according to the role they play (or they might play) towards the project (financing, active involvement, not opposing, etc.).
- The views on the different types of approaches and technologies regarding the implications to different stakeholders (cost, location, land demand).
- The interest to have additional water for a specific use.

4.3 ATTITUDES

The information people have and their interpretation (perception) shapes the attitudes they have towards projects. Therefore, introducing new approaches and technologies for sanitation requires alignment with current stakeholder's perception or an ethical intervention to change it. Behavioural changes can be promoted in a person-by-person approach or working with social groups using integrated approaches. The strategies can be focused on a specific project, or encompass broader concepts such as water reuse, sanitation or SDG 6 (WHO-UNICEF JMP, 2019). Behavioural change for sanitation programmes implies decision makers possess adequate leadership, and the involvement of all stakeholders including those from the government in order to achieve the following:

- understand prevailing sanitation and water reuse behaviours and their determinants[3], bearing in mind that determinants are specific to each population group as they have different perceptions, needs and opportunities for change.
- focus on the determinants of behaviours,
- carefully design the model of behavioural intervention based on the determinants (Conway *et al.*, 2023; WHO-UNICEF JMP, 2019).

4.3.1 Attitudes and behavioural change interventions

To build a new vision, interventions for behavioural change must focus on the risks and on all the types of benefits sanitation and water reuse bring, the advantages of reusing water and the conceptualisation of wastewater and by-products associated as a resource. To this end, it is highly recommended that behavioural change interventions have at their core the message that wastewater, sludge, biosolids and by-products are resources and not merely waste. The different stages to design behavioural change interventions are provided in Figure 4.2.

---

[3] Determinant is a factor which affects decisively the nature or outcome of something.

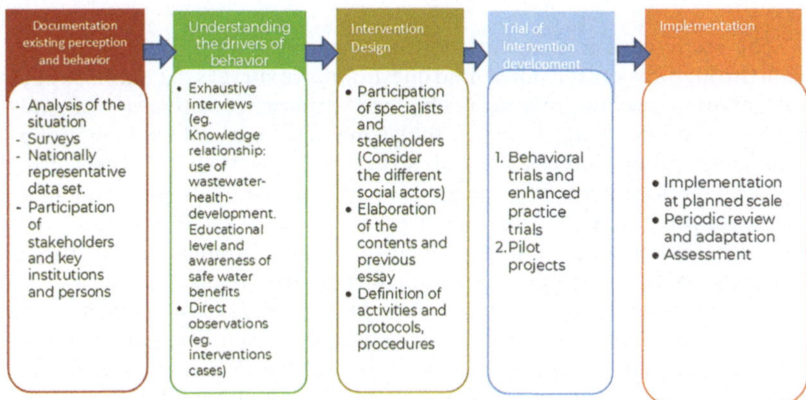

Figure 4.2 Steps to design a behavioural change strategy: new vision of sanitation and wastewater use (*source*: adapted from WHO, 2019)

Interventions to change perception and behaviours are not easy, fast or cheap (Box 4.3). Furthermore, literature shows that it requires highly specialised people. Aspects such as using native language, understanding past interventions and understanding social and political conditions are key. The participation of experts who understand the local cultural conditions and have sensibility is important. In addition, for rural and low-income regions, the approaches to change perception in most cases have succeed only if they are implemented in communities that are socially integrated (Kar & Chambers, 2008), which is frequently not the case where mixtures of social classes or cultural groups exist (Box 4.4). Thus, it is hard to implement this type of practices in countries with low incomes, which are precisely the regions where sanitation is needed the most. Therefore, in most cases decision makers will need to act using adapted methods and common sense.

Box 4.3 Behavioural change strategies design for sanitation is a complex task

A study on the emptying behaviour of OSS was performed in Khulna city, Bangladesh with a total population of 31,883 inhabitants. First, the determinants for attitudes/behaviours were identified, and then management strategies to motivate society to implement emptying programmes were designed. The study was carried out in ward no. 9 (smallest administrative urban unit) comprising 80 households using septic tanks, aqua privies and pit latrines. All the systems for collection and delivering of faecal sludge for proper treatment were inappropriate. Frequently, there was effluent overflow and faecal sludge was discharged into water bodies or open drains, causing serious public health and environmental problems.

The study used a mixed-method cross-sectional approach to collect data through structured household questionnaire surveys, face-to-face key informant interviews (informants had vast experience with practical and professional knowledge in faecal sludge management and OSS emptying practices), group interviews and structured observations. Literature and official documents of relevant governmental and non-governmental organisations (NGOs) were also used as a source of information. All research instruments were translated into Bengali language and three research assistants helped translating during interviews.

Emptying services description. The OSS emptying is performed by workers known as 'sweepers'. Irrespective of economic situation, more than half of the households either carried out reactive/emergency emptying or did not empty at all, violating the Bangladesh National Building Code which states an emptying period of 6–12 months. Emergency emptying was performed by contacting manual or mechanical/motorised sweepers when the system overflew or stopped working. In most cases, emptying was performed manually without safety precautions, appropriate personal protective equipment and working tools. There are approximately 150–200 active informal manual sweepers with a capacity to empty around 68.7 m^3 faecal sludge per day. For motorised sweepers, there were only two formally established service operators (the Conservancy Department of Khulna City Corporation and the Community Development Committee). These services work mostly for commercial, public and private organisations and for some households with spacious access to road connections. Mechanical/motorised emptying services were challenged by blockage of hosepipes and pumps and many narrow roads that made accessing individual households difficult.

Perception determinants, cues and strengtheners. Owners and/ or users of OSS based their emptying behaviours on their perceptions rather than on technical information or regulatory directives. To design a strategy to change the behaviour, the following four perception/ behaviour determinants were identified: (a) risks/hazards of emptying; (b) desired emptying behaviour; (c) emptying norms and (d) emptying ability. In addition, six cues were identified as potential triggers for stimulating emptying attitudes. These were: (i) past emptying behaviour; (ii) sanitation belief system; (iii) consequences of previous emptying experience; (iv) sanitation norms; (v) sanitation situational factors and (vi) confidence in ability to empty facility effectively. Information and cognitive processes were identified as perception strengtheners.

Perception of risks. Around 70% of the surveyed population did not perceive any negative consequence from unsafe and reactive emptying. Meanwhile, 88% perceived that their family members would face some risks of being sick from exposure to faecal sludge overflowing. However, they did not consider that there was much risk to public health. Only 23% expressed gross disgust for effluents from OSS systems in the drains and/or environment, considering this waste as 'not as bad as the solids' (faecal sludge).

Perception of the emptying process. Fifty-five per cent of the respondents perceived that engaging in scheduled (timely and safe) emptying was a tiresome process, 51% considered the activity as a costly venture and 88% perceive the practice would have a positive impact on the community.

Perception of norms. Sixty-one per cent of sanitation professionals and key informants knew that there were norms that prohibited discharging faecal matter into drains; however 42% perceived that its enforcement was weak. Concerning community members, only 6% were aware of such norms and 10% of them believed that enforcement and monitoring were adequate.

Perception of emptying ability. Nearly half of the households' respondents (56%) perceived that they did not have the required knowledge to determine when emptying was necessary, 60% believe that their knowledge on emptying facilities was adequate. Despite 80% thinking that scheduling was important, only 47% were willing to put an emptying schedule in practice. From the key informant group, although 70% were aware of the negative impacts of faecal sludge overflow on health, most people still chose to practice reactive/emergency emptying. In all cases, income restrictions were identified as a factor to delay emptying.

Models to change perception. On the basis of the findings, four scenarios were developed depending on the perception determinant to be used as a trigger for the emptying process: (a) scenario I: risks; (b) scenario II: desired emptying management; (c) scenario III: emptying norms and (d) scenario IV: emptying ability. Technical recommendations to facilitate the process were also elaborated based on the surveys' results, these were:

(i)   ensuring the availability and affordability of scheduled emptying services with ease of access to all strata of the community members;
(ii) review and update of relevant legislations, regulations, guidelines and standards for faecal sludge management;
(iii) strengthening compliance enforcement mechanisms from the government and
(iv) improving OSS infrastructure and ensuring that it complies with the provisions of the relevant codes and regulations.

In addition, techniques for controlling and directing perceptions were provided. These were: priming of desired perceptions; engaging lead users in awareness raising programmes; implementing reward systems using incentives for the community; using reference groups to influence perceptions; evaluating and addressing the feedback and providing information and education.

Source: with information from Cookey *et al.* (2020).

Box 4.4 Perception and behavioural change intervention in rural Odisha, India

In Odisha, India, studies on perception and behavioural change were performed during 2018 within the context of a larger project to increase basic sanitation coverage in rural villages. The purpose was to understand why people preferred to practice open defaecation instead of using the latrines that were going to be installed as part of the initiative launched by the Prime Minister Narendra Modi to make India open defaecation free by October 2019.

Odisha state had the second-lowest latrine coverage level (29%) in 2016. Since 2011 various government sanitation campaigns were rolled out in the region. During these campaigns government officials had forced villagers to stop open defaecation and used coercive tactics authorised by local government officials including harassment, public humiliation, fines and the threat or actual loss of public benefits.

Perception

According to surveys, individual and household sanitation was important in all 36 villages. People were interested in keeping their households clean, but not their village. Villagers practiced open defaecation in fields and threw waste (bags, plastic bottles, materials used to clean child faeces, etc.) outside the household compound into the open peripheral spaces. This is because sanitation is considered an individual and household-level priority, but not a common activity, requiring collective action.

The barriers to use latrines extended beyond access or ownership to include poor latrine design and construction, lack of water availability, fear of pit filling and the need to empty them, preference and perceived benefits of open defaecation and gender normative perceptions that latrines are only meant for women. The villagers doubted their village would be able to improve sanitation and latrine use due to conflicting interests across village divisions and a perceived lack of resources and authority to do so.

Need for outsider involvement in village sanitation. Villagers felt that their ability to influence others to stop open defaecation was limited because they did not have the necessary authority, especially when they lacked the resources to provide non-owners with facilities. Women experienced additional challenges in promoting latrine use due to gender norms, as they have no authority with men. Thus, they expected outsiders, such as the government, NGOs and contractors, to support village sanitation needs, including latrine construction, sanitation awareness campaigns and latrine use enforcement. However, outsider entities were also viewed as not reliable and frequently being dishonest, providing substandard latrines, and offering things which people sometimes found pointless, such as awareness and behavioural change programmes.

Participants on the surveys perceived awareness activities as repetitive and ineffective as long as government subsidies for sanitation and water for its functioning were not provided.

Village divisions. Sociocultural and geographic divisions existed in the villages with clear wealth disparities between hamlets (village sections), where members of different castes lived. They noted how these divisions obstructed change and development, as they were unable to reach consensus and support community improvement initiatives. In addition to this, they knew that leadership corruption and favouritism towards some social classes hindered progress. The differences between groups often escalated to violent conflicts. In occasions, the most deprived social groups were requested to be paid to support any activity that required an additional contribution from them.

Intervention to change perception and attitudes
In 2018, while the increasing sanitation coverage programme was being delivered, an intervention project was implemented for about 1 year in 36 villages of the Puri district. The implementation was performed by a local NGO Rural Welfare Institute (RWI) with support from Emory University. The project was called Sundara Grama ('Beautiful Village'). The intervention addressed six barriers to latrine use: (1) non-functional latrines; (2) limited practical knowledge on how to use a latrine and empty its pit; (3) preference for open defaecation due to attitudinal and sociocultural factors; (4) limited recognition of the health and non-health benefits of latrine use; (5) lack of infrastructure to aid safe disposal of child faeces and (6) limited knowledge regarding child faeces disposal.
To ensure all village members – men, women, children – were reached, eight methods were used consisting of a combination of community, group and household level approaches. The main results according to the type of intervention were as follows:

(1) *Pall* (traditional theatre representation which includes skits, songs and witty poetry). This form of intervention reached a large audience, was positively received and helped revitalise a traditional form of art providing messages on sanitation.
(2) *Transect walk*. The activity consisted of an early morning walk through the village and to open defaecation fields. Participants placed coloured powder on faeces. Many village members refused to participate, while others expressed anger, irritation, disgust and shame towards the act. In some cases, the project mobilisers were scolded for leading such an activity. The focus was centred on the shame it brought when the dirtiness of their villages was exposed to outsiders or when someone was caught in the act of open defaecation. The output of the intervention measure was not conclusive; some participants reported that it immediately led a

reduction in open defaecation, while others said any impact was short-lived or dependent on the individual.

(3) *Community meetings.* In these meetings, sanitation problems were discussed, and an action plan to address them was drafted. The identification of households where all members used a latrine all the time (positive deviants) were not appreciated by the community. Meetings were frequently disrupted by participants voicing their frustration at the poor quality of their government-provided latrines or not having received their latrine subsidy. Additionally, the differences among villagers were more noticeable. Lower-caste groups were forced to sit in a separate area or were not invited.

(4) *Community wall painting.* The painting showed both the action plan and a map of the village that indicated which households were positive deviants, identified in the community meetings.

(5) *Meetings for mothers of children under 5 years old.* The purpose of these gatherings was to provide action knowledge and hardware (potties and scoops) to aid safe child faeces disposal. The attendance was low, due to women being busy with other tasks and restrictive social norms around women participation in public meetings. The main attraction of the sessions was the potties, scoops and nappies that were provided.

(6) *Household visits.* The activities included either provision of a celebratory poster to positive-deviant households or household visits with non-users to encourage commitment towards latrine use.

(7) *Latrine repairs.*

Costs and overall results. For the 33 villages the total cost of the interventions was of 36,172 USD, that is, 18.49 USD per latrine. The costs included the intervention activities and the latrine repairs, excluding training and overhead costs. The *palla* accounted for 43.6% of the total delivery cost; RWI staff salaries and transport stipends accounted for 43.5% and the activity materials accounted for 12.9%.

In general, the project was viewed positively by villagers and it was perceived that it increased awareness around the importance of latrine use and renewed interest in village cleanliness. However, participants expressed mixed feelings around the effectiveness of the intervention, as they believe improvements to sanitation facilities and water access were still required. In fact, The Sundara Grama intervention resulted in a 6.4% increase in latrine use and a 15.2% increase in safe child faeces disposal.

Challenges. The main challenges faced were conflicts with villagers around government latrine subsidies and construction quality, stakeholders' support and social dynamics related to caste, gender and age. In some occasions, caste divisions compelled to organise two separate *palla* performances and community meetings, or in the most extreme cases to the exclusion of a group.

For the mobilisers (animators) challenges were to travel to their assigned villages and being misconstrued as government officials. Finally, female mobilisers – many of whom were young and for whom this was their first job – reported being catcalled and shamed by community members as their presence defied social norms restricting the mobility of young women.

Source: with information from De Shay *et al.* (2020), Sclar *et al.* (2022).

4.3.2 Using legislation to guide public perception

Shaping and changing perception and attitudes can be performed through an adequate legislation framework and its enforcement, accompanied by communication, awareness raising and educational programmes. Legislation is useful to promote sanitation and reuse projects. For politicians, presidents, governors and majors having an approved legislation is a good incentive to perform projects to fulfil the law.

4.4 COMMUNICATION AND AWARENESS RAISING CAMPAIGNS

As previously noted, social and stakeholders' acceptance and participation on projects and programmes depends on knowledge availability (Liberath Msaki, *et al.*, 2022; Saad *et al.*, 2017). Therefore, it is important for policy and decision makers to manage communication programmes. Most sanitation policy and decision makers that face the challenge of implementing sanitation are located in middle- and low-income regions[4] within their country, thus they do not have the same financial and human resources support than national programmes have. Thus, it is important for communication activities to be implemented without high-cost programmes. Training of sanitation leaders and those working around them must include highlighting the importance of diplomacy – that they must convey the right messages to specific communities and under appropriate circumstances. Having access to public information is now a human right and the public must be informed on projects, laws and programmes. Furthermore, it has become common for the public to request additional and specific information and even to get verbal or signed commitments with authorities regarding their initiatives on public services (Sanitation and water for all, 2021).

4.4.1 Managing communication campaigns

Managing communication campaigns' information is necessary to promote learning, inform people and reach a common understanding among all

---

[4] Where sanitation is needed the most, frequently the economic situation is not good. But, this could change if sanitation is coupled with water reuse projects.

stakeholders. As different people learn and communicate differently, it is important to carefully select the information to be used and to analyse the approach that will be implemented to disseminate it. To gain public participation it is important to constantly raise awareness on general subjects such as basic sanitation, wastewater treatment and water reuse. Sometimes it may be relevant to cover broader aspects considering the integrated management of water resources, health risks, economic benefits and how other parts of the world deal with the same challenges (BIO by Deloitte, 2015).

Public campaigns should also be performed on specific projects before their approval and implementation and when projects have been implemented. The latter in order to educate users on tasks they must perform, such as the timely and proper emptying of OSS systems or continuously performing complementary practices on reuse projects. To ensure the engagement of all the community, campaigns must be multi-cultural, multi-lingual and multi-ethnic. When it comes to the design of the communication programme, four aspects need to be considered: (a) the type of audience, (b) the messages to be conveyed; (c) the context and (d) the method that will be used for communication (Table 4.3).

Table 4.3 Aspects to be considered when designing communication campaigns.

Concerning the audience
- Every audience is different, and in general always avoid assuming knowledge or priorities of your audience.
- Attract different groups to your messages by tailoring them based on their top priorities (not yours), and perceived risks.
- Understand what the mutual benefits are for any non-sector actor.
- Understand your audience and consider that there might areas that may not be relevant to them at the moment. Be concise and to the point.
- Shape specific messages to target women, youth and ethnicity.

Concerning the content
- Communicate few core ideas using short messages that can be easily retained. Providing too much information distracts the audience and leads to confusion.
- Avoid providing generic information that anyone could give.
- Ground your messages in sound scientific and technical information. It suffices to provide mistaken information once to have an entire speech or person discredited.
- Sanitation and water reuse are two highly technical subjects, do not aim to educate people during communication campaigns or through messages, keep the focus on what behaviour you need from society/stakeholders/politicians when designing your programme.

Concerning the context
- Use local and accessible language.
- Define terms as much as needed.

Concerning the method
- There are different methods to present and communicate, consult specialised literature.
- If the subject is too technical, preferably use experts to communicate.

Source: partly, with information from BIO by Deloitte (2015); De Shay *et al.* (2020) and Sanitation and water for all (2021).

Besides the people from the community to which the communication and awareness raising programme will be addressed, there are two groups of interest which are important to reach: politicians, high-level decision makers and donors, as well as, policy and decision makers from other sectors. Tables 4.4 and 4.5 show recommendation to target these groups of people.

4.4.2 Awareness raising and communication for water reuse

Water reuse is a specific topic under sanitation that deserve special awareness raising, and communication programmes for three reasons: the language to be used, the ample set of actual and potential stakeholders that are part of other sectors and the need to implement intervention measures for water reuse in local contexts.

For water reuse, the language used to communicate is critical to influence public perception. The mental images or associations to the words 'wastewater', 'reclaimed water' or 'reused water' are different for each word (US-EPA, 2012a). In some places, the legal classification of 'toilets flushing' as part of drinking water uses automatically makes people reject the reuse of reclaimed water (Liberath Msaki *et al.*, 2022). To produce a 'paradigm shift' to perceive wastewater as a resource the language needs to be carefully selected (Rodríguez *et al.*, 2020).

Water reuse is a multi-sectoral group task. Different sectors need to cooperate and coordinate with different ministries and sectors at the national, regional and local levels. Among the social groups of interest, there are public agencies (notably those specialised in health, industrial and municipal development, agriculture and energy), property owners, special interest groups, customers, potential customers, academia with diverse specialities and the community surrounding the water reuse programme (Esquivel, 2015). Communication and awareness raising campaigns need to target all these different groups with specific messages.

In low-income regions with limited treatment capacity, raw or partially treated wastewater is often discharged into water bodies and then captured and used for informal irrigation. Under these conditions, the cultural and social challenges of awareness raising and communication campaigns are not to introduce water reuse as a practice for the community but to show risks and ways to control them in different ways that could be considered non-conventional (Scott & Becken, 2010) (Box 4.5).

4.5 EDUCATION

Education is a tool to provide knowledge that also changes perception and attitudes; however, educational programmes are not managed by sanitation policy and decision makers but by schools and universities. Education is a way of raising awareness on the importance of preserving water quality and the proper treatment and disposal of wastewater. Furthermore, it is a way to prepare people who may later contribute to communicate, change perception and attitudes towards sanitation. Educational programmes can

Table 4.4 Considerations to prepare messages on sanitation for politicians, high-level decision makers and donors based on their perceived risks and priorities.

Perceived Risks /Audience Priority	Messaging
Sanitation is perceived as costly.	• Show the cost–benefit relationship (or 'return on investment'), highlighting links to specific areas of interest for other sectors or ministries. For instance, contribution to the economic growth, health benefits and mitigation of greenhouse gas (GHG) emissions.
Inaction can cause political or economic drawback or delaying the fulfilment of other sectors/persons priorities.	• Underline the risks that poor sanitation means to other sectors' wider ambitions.
Supporting sanitation is not part of people's demands and priorities.	• Show that WASH is something that communities desire and demand as it contributes to improving health, education, gender equity and economic conditions. • Frame sanitation as aspirational and illustrate when improved how it can change communities' and families' lives for the better. • Keep in mind that the public perceives sanitation to be a government task.
Human security and supporting vulnerable groups are relevant.	• Show how sanitation contributes to increasing security and especially help vulnerable groups, such as women, Indigenous communities and deprived people. • Present sanitation as a way to contribute to addressing inequity and increasing human security.
Sanitation is an activity in which it is not easy to establish a political legacy.	• Explore how sanitation can be part of political legacy, to local politicians by drawing on and illustrating its long-term benefits. • Show that sanitation is a society (voters) demand. • Develop arguments to convince politicians of the importance of publicly discussing sanitation. • Understand and engage politically strategic groups that may influence leaders' perceptions of areas of electoral and political importance.
Sanitation programmes, projects and results are not aligned to political cycles.	• Align sanitation communications to political cycles (3–6-year time frames). • Consider framing your messages for politicians to be able to see results in the short, middle and long terms. • Prepare to address both inexperienced ministers and seasoned decision makers.
Sanitation is not part of the political or wider agendas.	• Frame sanitation in the context of wider political and global agendas, aligning ideas to politicians' priorities. • Political leaders must share their attention, funds and time between several priority areas. Compared to other priorities sanitation might not be at the top. They need to see how sanitation relates to their priorities to invest in it. • Convey the message that sanitation is a society demand as they move to better economic conditions.

Source: partly with information from Sanitation and water for all (2021).

Table 4.5 Recommendations to prepare messages to communicate with different sectors on sanitation.

Sector	Message Content
All sectors	Focus communications on the sector' priorities and perception. Leave for later educational communication to highlight the advantages of sanitation for sectors less familiar with the topic.
Economic development	Value robust, recent hard data that show the fiscal benefits of investing in sanitation. Focus on the risk of inaction (financial costs) of improving sanitation services. Show how improved sanitation services can increase economic and workforce productivity.
Health	Emphasise the short-term risks of poor access to sanitation services on health. Present the relationship between sanitation and health not as a scientific discussion but as a community aspiration (living better and healthier). Show the relationship between clean water, sanitation, hygiene and clean surroundings with better nutrition and health.
Education	Highlight the risks of poor sanitation services on education. Show that quality education cannot be provided without good quality sanitation services. Focus on aspirational messages showing how sanitation and WASH link to better education and therefore to higher household incomes. Present cost–benefit analysis showing that sanitation and WASH investment improves communities and education. Communicate the links between taxes and public services as a contribution to the country's development.
Human rights	Illustrate how sanitation helps create safer societies, particularly for marginalised groups, women and girls. Refer to WASH as a way to deliver the health and well-being objectives of the SDGs. Highlight long-term risks due to poor sanitation services. Frame sanitation as a security issue.
Humanitarian response	Emphasise the short-term risks of poor access to sanitation services, and its connection with safety. Focus on water as a priority need, and value hygiene and sanitation in crisis contexts. Highlight how sanitation protects marginalised people during catastrophes (more diseases in these contexts mean more suffering) and in the context of managing migration. Show that sanitation is cross-cutting to human rights, health and climate. Take a systems approach perspective, particularly to engage political leaders, considering the broader contexts of peace and security.
Climate change	Highlight the benefits of sanitation and water reuse on improving climate resilience and adaptation. Highlight how sanitation and water reuse can contribute to mitigate GHG emissions.

> ## Box 4.5 Implementation of unconventional options for safe water reuse in agriculture
>
> In Ghana, urban vegetable farming has relied on low-quality irrigation water for many years. Often, farmers have no choice but to use polluted irrigation water, which in most cases is more affordable, reliable and allows for year-round cultivation of vegetables. Risk assessments conducted in major cities in Ghana show high levels of faecal contamination in irrigation water and in vegetables grown with this water, which can lead to an annual loss of 12,000 disability-adjusted life years (Amoah *et al.*, 2005; Razak & Drechsel, 2010). To control the risks, the government followed the multi-barrier approach recommended by the WHO (2006). For example, farm-level water treatment was combined with good irrigation techniques, better management at markets and household vegetable washing for further cumulative reduction in contamination. To implement such a multi-barrier risk control strategy, a participatory approach using awareness raising and communication campaigns was adopted in which key stakeholders such as urban vegetable farmers, vegetable vendors, street-food vendors and local authorities (agriculture, health) were involved. To change the attitudes of stakeholders and adopt new practices economic or social incentives for behavioural change were identified, farmers were enabled to see and understand the 'invisible risk' caused by microbial contamination and sessions to exchange innovative empirical knowledge between key stakeholders and scientists were conducted. During the implementation of this participatory approach results and information gathered were used to make farmer-friendly extension and training materials. The materials were translated into different local languages, and presented in various forms such as illustrated flipcharts, books, radio and video and presented in field schools and farmers' markets. During the participatory approach and awareness raising campaigns local authorities and relevant government ministries were involved from the beginning, for instance the Ministry of Food and Agriculture and the food safety regulators. Additionally, the project on water reuse was linked to food security projects.
>
> *Source*: with information from Keraita and Drechsel (2012).

be prepared for the general public (Box 4.6) and for professionals with different education levels (technicians, bachelors, masters, PhDs) in schools and universities (Box 4.7). Given the magnitude of the effort required, it is critical to include sanitation in the curricula of universities, vocational schools and other specialised institutions, such as local public training centres. Peer learning and mentoring are also very effective for education on sanitation.

Box 4.6 NEWater Visitor Centre in Singapore

The Public Utilities Administration of Singapore (PUB) developed the 'ABC Waters Program' as part of a comprehensive approach to promote water reuse through awareness raising and education. As part of it, the 3P programme (for people, public and private sector) was implemented using community leaders, journalists, trade groups, government offices and the media to voice key messages and provide information. The NEWater Visitor Centre was built to offer public educational programmes and to disseminate information. It attracted 800,000 domestic and foreign visitors in the first initial 5 years. To minimise negative public perception and fears and stigma, PUB adapted technical information and terminology by using simple vocabulary; for example, the term 'wastewater' was replaced by 'used water', and instead of 'wastewater treatment plant' they used 'water recovery plant'. Information on water reuse is presented using simple diagrams and graphs, and fun tools to engage the community, for example, a video game called 'Save My Water'. The social acceptance of wastewater as a resource increased following these educational initiatives.

Source: with information from www.pub.gov.sg/ (PUB); https://www.pub.gov.sg/Public/Places-of-Interest/NEWater-Visitor-Centre.

Box 4.7 Education and awareness raising on sanitation and reuse in Latin American universities

Educational project to make children and young people aware of the youth–science–health–environment relationship.

In Latin America one of the causes of waterborne diseases and water pollution is that most of the population is unaware of the close relationship between water quality and human and environmental health. To address this issue two projects were conducted: one to educate children and young people on these aspects and another to train teachers, researchers and water science and technology students to better contribute to this awareness raising project.

The first project was carried out in Argentina, Brazil, Bolivia, Chile, Colombia and Uruguay by one or more universities in each country with the support from water and wastewater utilities, health ministries/secretaries, schools and colleges, municipal governments, NGOs, industry, banks, Pan American Health Organization and UNICEF. The project consisted in teaching children and teenagers how to test water quality in their surroundings, find pollution sources and understand the associated effects on public and environment health. The educational sessions were developed using different platforms for each university. To evaluate the quality of the water, field tests were conducted to observe the effects of toxins on plants (*Allium cepa* and *Lactuca sativa*) and animals (*Hydra attenuate*). Water quality analysis was performed during workshops at schools during which results were also discussed. In total, for the seven Latin American countries, around 1,200 children and young people were trained. The workshops were implemented for 10 years in Colombia, mainly for children displaced by violence.

Children and youth education and awareness-raising activities

Water risk identification (research)   *Risk minimisation (water treatment)*

The second project was conducted only in Peru to enhance the educational institutional capacity and to train teachers, researchers and students in water science and technology to better contribute to the awareness raising project.

Source: with information from Campos *et al.* (2001), Castro and Aurazo (2007), Pastor *et al.* (2011), Pastor and Miglio (2013).

doi: 10.2166/9781789064049_0109

Chapter 5

Sanitation costs and financing

KEY MESSAGES:

- Safely managed sanitation improves quality of life and saves lives. Sanitation is a low-regret climate adaptation strategy, contributes to the achievement of sustainable development goal SDG 2: Zero hunger, SDG 3: Good health and well-being, SDG 4: Quality of education and, given the differential impacts on women including needs for menstrual hygiene management, it is essential for achieving SDG 5: Gender equality. Considering all these benefits means sanitation always has a positive benefit/cost relationship, especially when incorporating social as well as economic perspectives.
- The cost associated with inadequate sanitation can achieve 7.2% of gross domestic product in some countries and has an estimated returns of 4–5 USD for every 1 USD invested. Notwithstanding, only 25% of all countries are on track to achieve their sanitation targets and investments for sanitation always lags behind that of drinking water.
- Within the sanitation sector, the investments made to manage sludge and faecal matter are well below those made to commission wastewater treatment plants and on-site sanitation systems.
- Households are by far the the largest sanitation funding source (covering approximately two-thirds of costs). This comes from both water tariffs and population self-supply.
- The costs for sanitation include all the elements of the sanitation chain, as well as human resources, capacity building and public participation. Financing them all is relevant, but it is also important to track which segments of the population are paying for what.
- The financing mechanisms that were used in the past to provide sanitation were not only insufficient but also inadequate to fund the varied set of tasks and groups that the entire sanitation chain implies. New mechanisms are available nowadays and their use depends on local conditions.

- Ensuring sanitation for all requires financial tracking to ensure that funds are reaching the targeted population.
- Having access to funds is not the only aspect necessary to effectively finance the sanitation chain. Proper and expedite procurement and expenditure mechanisms are needed too, of course accompanied by tools to ensure transparency and accountability.

5.1 WHY SANITATION IS IMPORTANT FOR THE ECONOMY

Ultimately, safely managed sanitation improves quality of life and saves lives. Sanitation is also a low-regret climate adaptation strategy (Caretta *et al.*, 2022), contributes to the achievement of sustainable development goal SDG 2: Zero hunger, SDG 3: Good health and well-being, SDG 4: Quality of education and, given the differential impacts on women including needs for menstrual hygiene management, it is essential for achieving SDG 5: Gender equality. However, only 25% of all countries are on track to achieve their sanitation targets, in part because approximately 80% do not have sufficient funding to meet their national targets (WHO, 2022). However, having funds to develop infrastructure is not enough, as better-performing countries are those that have a higher utilisation of domestic capital commitments, a good recovery from tariffs to support operations and maintenance costs, provide financing for human resources and build capacity on financing. This chapter provides data on the current situation in sanitation financing at the global level and analyses cost elements, financing mechanisms and the efficiency of expenditure. This information aims to assist to perform better financing for sanitation.

Sourcing the financing for the up-front capital costs of sanitation is one of the most significant challenges to scaling up to universal access. Operation and maintenance costs can also be significant, particularly in regions where energy costs are high or supply chains are fragmented. Further, many countries lack sufficient human resources capacity to construct and run sanitation facilities (GLAAS, 2022). Sanitation costs must also start to recognize, account for, and address hidden costs, such as greenhouse gas emissions.

5.2 CURRENT FINANCIAL SITUATION

Providing a global view of the financial sanitation situation is not an easy task because little information is available and the existing data depend on local conditions, the methodologies used to gather and analyse them and the scope of the economic analyses. Further, a lack of data on sanitation costs across the service delivery spectrum makes it difficult to account for all costs involved in sanitation service delivery and therefore to finance all aspects sustainably. Notwithstanding, the data gathered in the GLASS 2021/2022 country survey complemented the analysis performed by WHO (2022), which provides useful insights to understand the needs. The GLASS survey compiles data on water, sanitation and hygiene (WASH) from 121 countries and territories but not all countries responded to all or the same questions. Therefore, here instead of presenting figures, percentages are used as much as possible.

5.2.1 Funds needed to achieve SDG 6.2 and impact of the lack of sanitation on the economy

An estimated 105 billion USD is required to meet the sanitation component of SDG target 6.2: End Open Defecation and Provide Access to Sanitation and Hygiene from 2017 to 2030 (Hutton & Varughese, 2020). Breaking this down, basic sanitation[1] accounted for 34% and safely managed sanitation[2] accounted for the remaining 66%. To eliminate open defecation, the first-time capital costs are 1.5 billion USD annually with significantly greater capital replacement costs of 3.9 billion USD annually (Hutton & Varughese, 2020). From the total fund needed for sanitation, 70% is for basic sanitation in urban areas with the highest investment needed in sub-Saharan Africa (Hutton & Varughese, 2020).

Historically, the World Bank has assessed global national costs associated with inadequate sanitation to be around 260 billion USD per year (Hutton, 2013), with the regional distribution presented in Figure 5.1. National losses present significant variations among countries; in some they can attain

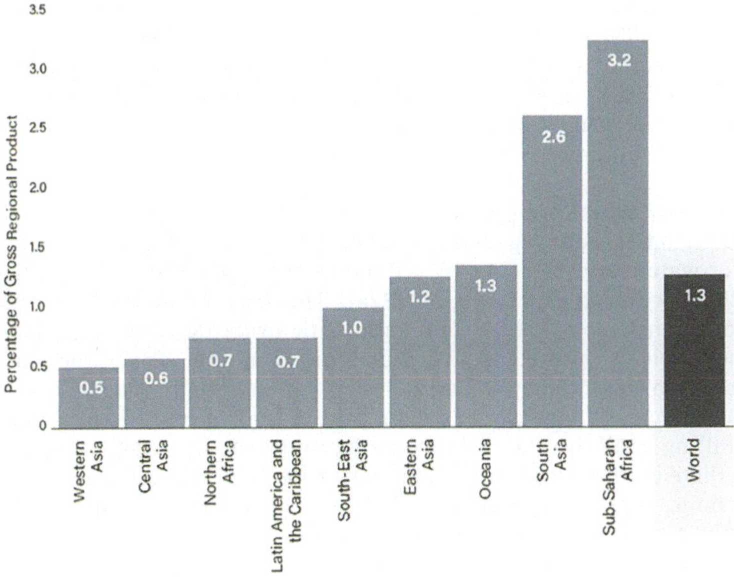

Figure 5.1 Economic losses in the Global South associated with inadequate sanitation by region, as a percentage of GDP in 2012 (*source*: UNICEF & WHO, 2020).

---

[1] For UNICEF basic sanitation facilities are improved sanitation facilities that are not shared with other households. They include flush/pour flush to piped sewer systems, septic tanks or pit latrines, ventilated improved pit latrines, composting toilets or pit latrines with slabs.

[2] Following the same logic as that for basic sanitation, in this case safely managed sanitation costs refer only to costs of safe excreta management; costs of latrine are not included.

7.2% of the gross domestic product (GDP) (World Bank, 2023a). Most losses are due to premature mortality, time lost while accessing sanitation, loss of productivity, additional costs for healthcare and the loss from potential tourism revenue (WSP, 2011). Considering these losses, financial savings outweigh the sanitation investment costs required, with a historic estimated return of 4.3–5.5 USD for every 1 USD invested (Hutton, 2013; UN, 2014; WWAP, 2017).

5.2.2 WASH global financing situation

According to the GLASS 2021/2022 report, the average annual total WASH[3] budgets have been increasing in most countries by a larger amount than the population growth. For example, they rose from 0.73% of the GDP in 2016/2017 to 1.10% of the GDP in 2021/2022 (WHO, 2022)[4], that is, approximately 50% increase whereas population growth rose by 5% in the same period. The budget of 71 countries amounted to 31 billion USD, equivalent to 12 USD per capita, a value that is only indicative as there are wide variations among countries. From the total budget for water services, drinking water amounted to 56% and sanitation to 44% whereas 62% was for urban areas and 38% for rural. Based exclusively on the sanitation budget, 56% is for urban areas and 44% for rural ones (WHO, 2022). However, budgets alone do not reflect availability of funds. Only 25% of the countries reported having sufficient funds[5] for WASH (i.e. 75% lack sufficient funding) (WHO, 2022).

5.2.3 Financial needs and sufficiency of funds for sanitation

Sanitation has always lagged behind drinking water, with shares from the total for water services being 46, 35 and 22% for 2016/2017, 2018/2019 and 2021/2022, respectively (WHO, 2022a). The lack of funds for sanitation is greater when compared to not only with drinking water but also with hygiene (Table 5.1 and Figure 5.2). For sanitation, only 22% of the reporting countries have sufficiency of funds for urban sanitation and 15% for rural ones, a percentage that decreased to 14% in both cases, when considering the fulfilling of all the national targets of sanitation. As a result of the lack of funds, there are limitations to capital expansion of services to reach unserved populations, deferred operation and maintenance that may ultimately lead to higher capital renewal needs in the future, and limited human resources to implement programmes and services (WHO, 2022). When only the funds needed to cover human resources are considered, 7% of the countries reported to have a formally approved policy and costed plan with sufficient funds for human and financial resources for urban sanitation and less than 3% for rural areas (WHO, 2022).

---

[3] Wash subsectors are drinking water, sanitation and hygiene.
[4] World population of 7 492, 7 578, 7 888 and 7 975 thousand million people for 2016, 2017, 2021 and 2022, respectively.
[5] Sufficient funding means having more than 75% of what is needed to implement WASH (Sanitation) plans or reach WASH (Sanitation) national targets.

Table 5.1 Comparison of sufficiency of WASH subsector funding to implement associated plans and achieve national targets.

WASH Area	Percentage of Countries with Costed Plans that Reported Sufficient Funding to Implement Plans	Percentage of Countries that Reported Sufficient Funding from all Sources to Achieve National Targets
Urban sanitation ($n = 100, 91$)	22	14
Rural sanitation ($n = 96, 90$)	15	14
Urban drinking water ($n = 97, 92$)	23	29
Rural drinking water ($n = 95, 89$)	23	25
Hand hygiene ($n = 83$)	NA	27
WASH in healthcare facilities ($n = 63, 84$)	32	25
WASH in schools ($n = 71, 82$)	35	23

Source: with data from WHO (2022).
NA, sufficiency of costed hygiene plans was not asked in the GLAAS 2021/2022 country survey.
n, number of reporting countries.

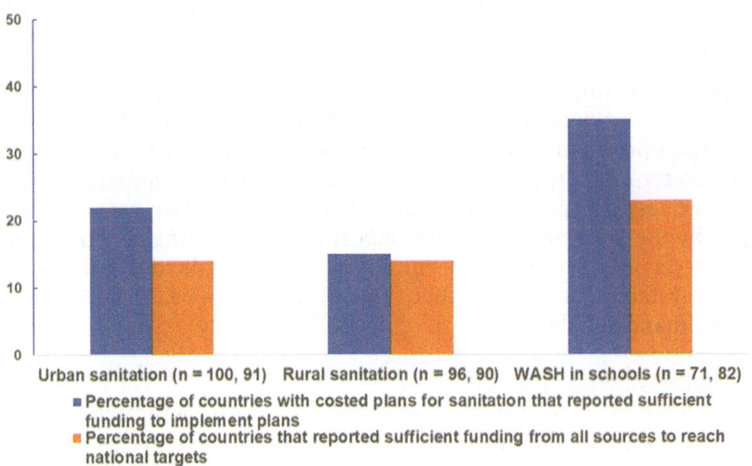

Figure 5.2 Sufficiency of funding to implement sanitation plans (*source*: with information from WHO, 2022).

5.2.4 Funding sources

Households, not governments, currently provide the largest proportion (approximately two-thirds) of funding for WASH (Figure 5.3). In the case of sanitation the funding is through connection tariffs and self-supplied services such as covering on-site sanitation emptying service fees and investing in toilets, on-site containments and, even, treatment technologies (WHO, 2022).

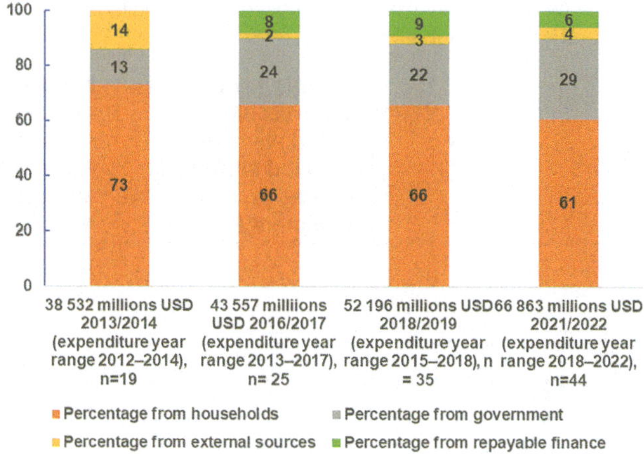

Figure 5.3 Breakdown of WASH funding sources from the last four GLAAS cycles for all responding countries[6] (*source*: with data from WHO, 2022).

5.3 SANITATION COSTS

Although it is recognised that those responsible (e.g. municipal governments) need to ensure that costs of sanitation services are as low as possible while maintaining standards of service and treatment, it is also admitted that a good and complete costing of the service (all along the sanitation chain) needs to be recognised and performed. As discussed in the following sections, costing for sanitation provision is different from that of other public services such as drinking water or energy (Mills *et al.*, 2020), because for sanitation there are several services and operational liabilities that are placed on users and are not provided by the government. This is important as budgeting constraints and a short-term vision for investments cannot be used to make decisions on which sanitation system to invest in; rather, it should be driven by the need to address long-term financial liabilities.

5.3.1 Sanitation cost components

There are several components that made up the costs for sanitation (Figure 5.4), including some that have been presented in current literature as externalities but as well, hidden costs. In any case, it is important to ensure as much as possible that there are sufficient funds for the proper construction, sustainable operation and maintenance, the rehabilitation, replacement and upgrading of facilities and to cover the needs for adaptation and mitigation of climate change. In some cases, too much emphasis is placed on promoting and financing new infrastructure, without sufficiently considering the life cycle of a

---

[6] Data include countries that provided total WASH expenditure data and information on household and government expenditures for each GLAAS cycle.

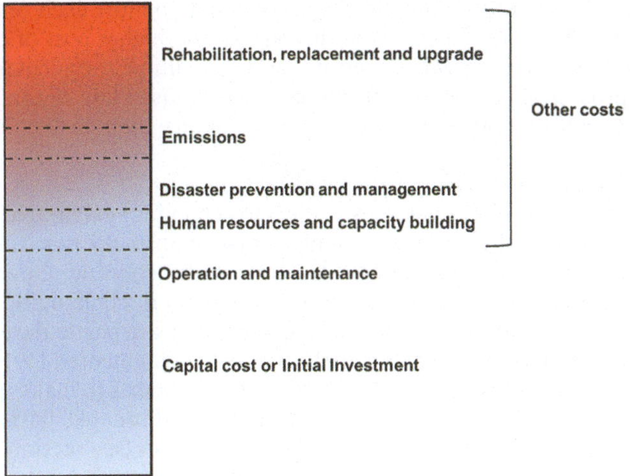

Figure 5.4 Components of total sanitation costs.

treatment system, the sustainability of the service, the real capacity of existing infrastructure and the maximisation of its use (UN-Habitat & WMO, 2021). This section will analyse the cost components for sanitation. These components apply to the entire set of elements that are part of the sanitation chain.

5.3.1.1 Capital costs (initial investment)

The investment and capital expenditure costs include the construction of sanitation facilities to expand coverage and improve service quality. The upgrading and the renewing of existing facilities are not, at least usually, considered under this. The type of sanitation option (decentralised or centralised) is critical in determining the sanitation costs (Box 5.1). In general, centralised systems are more costly

Box 5.1 Comparing costs among different sanitation approaches

A study conducted across five cities from Kenya (Kisumu, Malindi and Nakuru), Ghana (Kumasi) and Bangladesh (Rangpur) estimating the total financial requirements for achieving universal sanitation in the next 10 years showed that costs varied greatly between sanitation approaches. The study compared three to five sanitation approaches in each city: sewerage was the costliest approach (total financial requirements of 16–24 USD/person/year), followed by container-based sanitation (10–17 USD/person/year), on-site sanitation (2–14 USD/person/year) and mini-sewers connecting several toilets to communal septic tanks (3–5 USD/person/year).

Source: with information from Delaire *et al.* (2021).

to build than on-site ones; however the service and comfort each one provides to users is not the same. Some of the reasons for the higher cost of centralised systems are that sewers are very expensive, and in addition, they demand specific and costly process for construction. For both, centralised and decentralised, in most cases people do not have enough funds to finance their construction.

5.3.1.2 Operation and maintenance costs

Operation and maintenance costs are critical to achieving sustainable sanitation services as many of the sanitation-built infrastructures do not function due to a lack of maintenance or the cost to cover its functioning. Operation and maintenance costs, even if lower than investments, are challenging to afford, especially in low-resource settings; notably, when they are particularly high as a result of the energy costs or when supply chains are fragmented (WHO, 2022). In most of the cases, the approach taken by policy makers is to make users cover the operation and maintenance costs. This can be unaffordable, particularly for the poor, requiring subsidy or cross-subsidy programmes (see section 5.4.1.1).

An assessment of sanitation costs from multiple nation states across different urban sanitation solutions was used to calculate total annualised costs per household and per capita (Sainati *et al.*, 2020). Findings show that a focus on capital costs to the exclusion of operating costs can create a false understanding of costs associated with different sanitation solutions. For example, although capital costs of sewerage systems are higher per capita than faecal sludge management systems, operational liabilities are far lower for sewerage systems than for faecal sludge management systems that require road transportation of sludge (Sainati *et al.*, 2020). Furthermore, operational costs are far higher for container-based sanitation systems than for on-site systems (Sainati *et al.*, 2020). Sanitation storage systems require emptying, which, in turn, requires additional transportation infrastructure and off-site wastewater treatment. However, these types of sanitation solutions are also most likely to align with pro-poor financing strategies (Hutchings *et al.*, 2018).

5.3.1.3 Human resources, capacity building and awareness raising

Human resources and capacity building are part of the costs to be considered for both capital and operation and maintenance costs, all along the sanitation chain (Dickin *et al.*, 2020; Mills *et al.*, 2020). According to GLAAS 2021/2022, the lack of trained professionals 'is a major impediment to achieving safely managed sanitation'. The quadrupling of progress required to achieve SDG 6.2 necessitates an equivalent increase in trained workers for all facets of sanitation. However, almost none of the nation states reporting in GLAAS had formally approved policies with costed plans and sufficient human and financial resources (WHO, 2022). But the challenges do not stop there. Another challenge is retention of sanitation workers, particularly in rural, remote and indigenous communities where desired quality of life, isolation and pay inequities can all serve as barriers to retaining highly qualified people (GLAAS, 2022a; Murphy *et al.*, 2015; World Bank Group, 2019a).

Additionally, several case studies have identified that sensitisation is essential for sanitation investments (e.g. Post & Athreye, 2016), but it needs to be publicly funded to be delivered through civil society organisations. As an example, the breakdown costs for decentralised systems in Ghana and Ethiopia (Crocker *et al.*, 2017, 2021), are as follows: 17% for management, 46% for training and 37% for facilities. The breakdown for individual latrine costs consisted of 30% to cover local actors and community members' time, 52% to buy materials and 18% for external labour.

5.3.1.4 *Disaster risk prevention and management*
Unfortunately, experience shows that there is a need to include disaster risk reduction measures as part of sanitation costs, even without climate change scenarios. Contingency funds are needed to repair infrastructure and restart operations after extreme rain events, floods, tsunamis and earthquakes. These funds can come from any source including insurance agencies (Dickin *et al.*, 2020; Mills *et al.*, 2020). Impacts of climate change will increase these needs in many regions. However, only one in five nation states have scaled approaches for climate preparedness in the water and sanitation sectors, whereas one-quarter have pilot or demonstration sites established (GLAAS, 2022b).

5.3.1.5 *Greenhouse gas emissions*
Centralised wastewater treatment plants (WWTPs) account for approximately 3% of global energy use. Depending on how this energy is generated, these plants contribute varying amounts of greenhouse gas (GHG) emissions over and above direct emissions associated with the breakdown of organic matter in the treatment process (Dickin *et al.*, 2020). This decomposition also occurs in the environment when no treated options are provided. As sanitation services such as pit latrines are provided to the 2.3 billion people who still lack access to basic sanitation, GHG emissions could more than double (Dickin *et al.*, 2020). Investment in technology to minimise such emissions will be required. Energy and nutrient recovery could reduce the need for other forms of energy consumption (for example, commercial fertilizer production).

5.3.1.6 *Technology development*
Designing and testing appropriate technology to ensure that it meets user needs and regulatory requirements incurs costs. Frequently, technology research and development is covered by the research and academic sector, which might include public or private funds, and international aid. On-site pilot testing costs are normally covered by sanitation utilities or NGOs in the case of household sanitation solutions.

5.3.1.7 *Rehabilitation, replacement, expansion and upgrades*
The cost for rehabilitation, replacement, expansion and upgrades can be more difficult to finance. Although centralised sewerage systems have typically been the choice in large urban centres, many are ageing, particularly in high-income countries, creating a significant financial liability. Back in 2005, typical replacement costs were estimated to be 2,600 USD per capita for high-population countries and 4,800 USD per capita for those with lower populations

(Maurer *et al.*, 2005). In recognition of this liability, many urban management districts around the world have undertaken asset replacement calculations. Although a current global figure for wastewater infrastructure replacement is not available, replacement costs could triple within 5 years given a construction inflation of 5% (Sunshine Coast Regional District, 2019).

5.3.2 Urban, small and rural infrastructure costs

Urban centres represent significant numbers of people in relatively small geographic areas. Although this can represent challenges in installing or expanding centralised systems through areas of pre-existing dense buildings, the cost per capita investments decrease with an increasing numbers of people served and revenue from direct fees and taxes increases with increasing population (Schuster-Wallace & Dickson, 2017). Conversely, the investment cost of centralised sewerage systems is far higher than many distributed or household-level solutions. However, development bank investments in centralised sewerage systems were 20 times greater than investments in more localised faecal sludge management systems between 2010 and 2017 (Hutchings *et al.*, 2018).

Rural and remote communities typically have small populations, less secure supply chains and greater expenses for transportation of goods and services (Schuster-Wallace & Dickson, 2017). In addition, they lag behind urban areas regarding their water supply and most of their sanitation infrastructure. Centralised sewerage systems are beyond the resource capacity (financial and human) of these and individual and small groups of households. Furthermore, access to land in some areas of cities, such as peri-urban areas or informal settlements, is a significant challenge to increasing urban sewerage services. As such, self-supply options (with or without supporting finance mechanisms) are typical, including shared (private or public) or stand-alone latrines with or without on-site wastewater treatment (Hutchings *et al.*, 2018).

This has two key consequences. The first is that the responsibility for operation, maintenance and replacement lies entirely with the household. The second is that not all self-supply options meet the criteria of improved access to sanitation defined by UN Water's Joint Monitoring Program led by WHO and UNICEF.

5.3.3 Local variation of costs

Cost cannot be generalised and needs to be determined on a case-by-case basis because they are highly dependent on local conditions. For example, field data from Ghana and Ethiopia for community-led total sanitation showed that initial average costs were 30 USD per capita in Ghana versus 14 USD in Ethiopia. Further costs for Ghana rose by 52 USD whereas for Ethiopia they rose just 5 USD (Crocker *et al.*, 2017, 2020). This high variability renders the estimation of global, regional and national costs a difficult task. Furthermore, these costs are not useful in defining local programme costs, although they do provide some guidance particularly for external investors.

5.3.4 Sanitation delivery chain costs

The complete life cycle cost of sanitation provision, including externalities, needs to be accounted for (Mills *et al.*, 2020). However, a detailed breakdown of costs along the supply and service chains is not yet available. Furthermore, because the entire sanitation chain consists of steps performed by different institutions, organisations and individuals it is difficult to identify a single agency that could be in charge of undertaking this effort.

5.3.5 Distribution of costs (who is paying what and how much)

As Mills *et al.* (2020) concluded, 'analysis of cost effectiveness against consistent service objectives will permit improved comparison of the mix of sanitation options likely to be appropriate to different contexts across a city. This will create an opportunity to then separately consider how costs may be fairly distributed across different actors'. Such an approach is critical to ensure affordability and equity in providing sanitation, notably because of the large sanitation market potential. Just as an example, the sanitation market potential in Kenya is estimated to be around 6.2 billion USD (Toilet Board Coalition, 2020) and 148 billion USD in India (Toilet Board Coalition, 2020) by 2030.

At present, it is clear that cost distribution is very different for conventional wastewater treatment and on-site sanitation systems. Using data from Dakkar, Senegal, Dodane *et al.* (2012) compared (a) an activated sludge and sludge management system with (b) an on-site sanitation system including septic tanks, collection and transport trucks and drying beds for the latrine content treatment. The annualised per capita capital cost for the conventional system (a) was ten times higher than the on-site system (b), while the annual operating cost was 1.5 times higher. The combined annual capital and operating cost per capita for (a) was five times higher (54.64 USD) than (b) (11.63 USD) (Figure 5.5). In addition,

COSTS, USD/year.capita	(a) Conventional	(b) OSS
Total	54.64	11.63
Utility	52.63	1.86
Household	2	9.74
Collection and Transport	0	0.02
End user	0.01	0.01

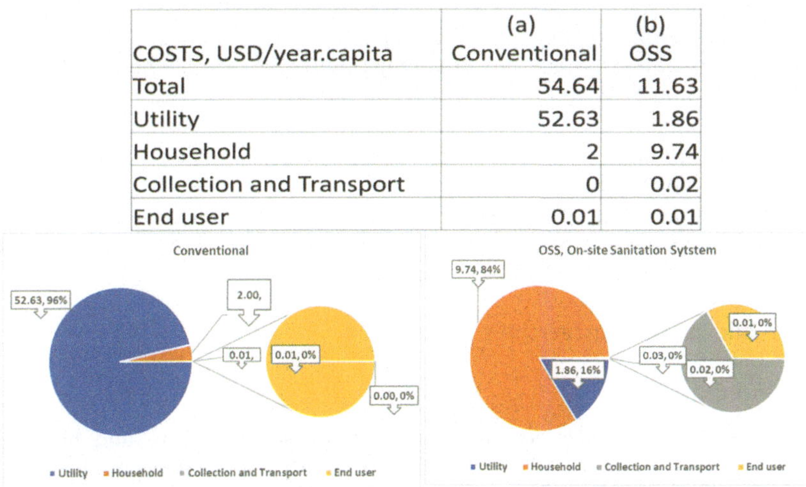

Figure 5.5 Costs and percentage they represent for conventional systems and OSS (on-site sanitation system) in Dakar, Senegal, with data from capital and operating costs of full-scale faecal sludge management and wastewater treatment systems in Dakar, Senegal (*source*: with information from Dodane *et al.*, 2012).

option (a) is not only 40 times more expensive than (b) but also costs are almost all borne by utilities while for option (b) costs are borne by the households.

5.4 FINANCING

Sanitation financing refers to the mechanisms and strategies used to secure funding for infrastructure and services. It involves identifying, mobilising and allocating financial resources to support the planning, construction, operation and maintenance of sanitation facilities, such as toilets, sewage systems and wastewater/faecal sludge treatment plants. Effective sanitation financing is crucial for achieving the goals and targets for universal sanitation. Sourcing the financing for upfront capital costs of the entire sanitation chain is one of the most significant challenges to scaling up for universal access (GLAAS, 2022a).

5.4.1 Options for financing

There are different sources and mechanisms for financing: those that have been commonly used in the past (and continue to be used), which we refer to here as 'conventional', and those that have less frequently been used, or are simply new, which we refer to as 'non-conventional'. Regardless of the source of financing, there are only a few mechanisms for financing. They are transfers, taxes, . tariffs (or fees), and individual investments (whether in capital or supplies and maintenance) (Danert and Hutton, 2020). Higher use of domestic capital and tariffs for operation and maintenance along with strong regulatory authorities and sufficient human and financial resources have been demonstrated to perform better with respect to increasing sanitation access (GLAAS, 2022).

5.4.1.1 Conventional options

The conventional mechanisms for financing (Figure 5.6) are (Danert & Hutton, 2020; UN-Habitat & WMO, 2021; UN-Water, 2024):

- Public funding,
- Households funding,
- International support via national or local governments,
- Cross-subsidies,
- Development aid, via national and local governments,
- Private,
- Public–private partnerships (PPPs).

Public funding

As part of the realisation of the human right to sanitation, and because of its impact on human and environmental health and economic and social development, governments play a crucial role in financing sanitation projects. In fact, government participation in financing is the only way to ensure that nobody is left behind and that needs for vulnerable populations are addressed. But the government's role in financing does not stop merely with provisions of

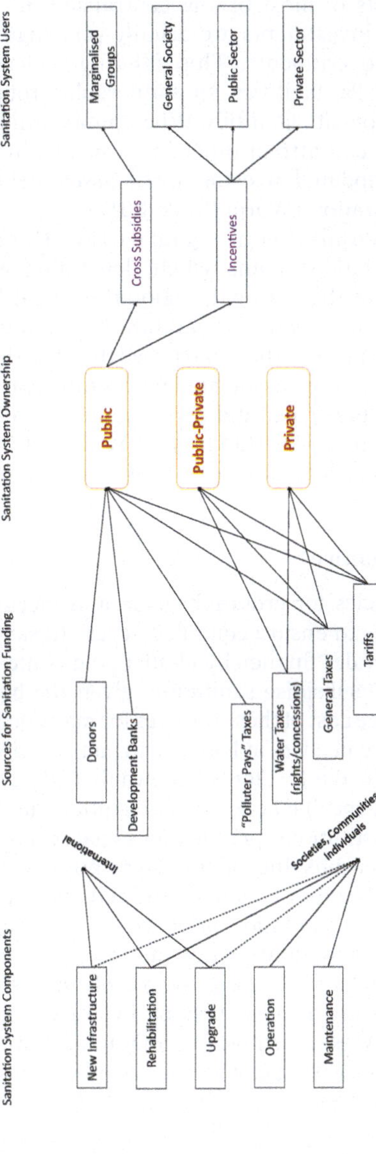

Figure 5.6 Conventional options that are used to finance sanitation.

funds. They have the unavoidable responsibilities of regulation, oversight and financial stimuli for non-government investment mechanisms, and must ensure that the entire sanitation chain has access to appropriate financing.

On the investment side, governments (national and regional levels) are needed to cover the costs of most of any centralised infrastructure, at least initially. This is because investments are usually very high and almost always unaffordable by local governments. This funding is used mostly to build sewerage systems, WWTPs, facilities for vulnerable groups and facilities to treat the containment of on-site facilities. When public funds are not enough to cover all the needs, they can attract other investments, including commercial financing, but may need updated investment risk assessments and clear policies and programmes for sanitation (World Bank, 2017).

Public funds can include ministerial, regional or local funds. Public funds can also include international aid or loans which, depending on the legal national framework, can be held at the national, regional or local level. Governments can also establish dedicated funds for sanitation or allocate resources from existing infrastructure funds. For these purposes they use the funds they collect from the right to use water, taxes associated with sanitation such as those using the principle of 'who pollutes pays' and taxes applied in general to individuals and business in other sectors. UN-Habitat and WMO (2021) assessed that the principle of 'the polluters pays' has been useful to put WWTPs in place in several countries.

Cross-subsidies and incentives

Governments define policies for cross-subsidies[7] and incentives for sanitation. This approach contributes to ensure equitable access to sanitation services and allows its sustainability under financially challenging contexts. Even if the goal of a government is not to subsidise sanitation, given the high investments and operation and maintenance cost of these services for many deprived communities, it is clear that subsidies are needed, at least during a transition period.

According to the World Bank (Flores Uijtewaal *et al.*, 2018), 'incentives' are created for the delivery of specific actions and resulting outcomes, from regional and local governments, institutions, private and social sectors who are motivated by political or economic reasons. Incentives can be positive or negative (perverse incentives) depending on the type of relations and actions they promote. Incentives can promote the funding of sanitation projects. For instance, legal declaration of the universal right to sanitation creates the incentives for municipalities to fund this area. The tax related to the principle of 'the polluter pays' and the establishment of subsidies are also example of incentives. In fact, many authors consider subsidies as a perverse incentive for sanitation. However, so far, the use of incentives to improve the provision of water services has not solved the problems, and services are becoming poorer worldwide (GWP, 2000, 2008).

---

[7] Cross-subsidisation involves using revenue from profitable sanitation services in affluent areas (or other sectors such as water, education or health) to subsidise services in low-income or underserved areas.

Household funding or self-supply

As mentioned, it is households, not governments, who currently provide the largest proportion of funding for sanitation and one of the ways in which they do this is through self-supply schemes. Self-supply funding includes all the expenditure made by the community or private persons to install and operate sanitation infrastructure. Unfortunately, most of the available information on households funding is difficult to trace due to lack of accountability for monitoring. Therefore, the out-of-pocket household contributions are underreported, and the available information is mainly estimated (WHO, 2022). Policy makers are using self-supply as part of a non-conventional approach to improve sanitation coverage. Notwithstanding, self-supply can perpetuate inequities, as, according to Augsburg and Rodríguez-Lesmes (2020), the highest income quartile represented those most likely to purchase a latrine, whereas the lowest income quartile was most likely to use informal loans or government subsidies.

Tariffs

Tariffs and user fees are charges set by the organisations running the services or by independent entities. They are set regardless of whether sanitation facilities are run privately, publicly or by the community. The economic resources from tariffs may go via the government or directly via sanitation utilities to provide the service. Implementing user fees or tariffs for sanitation services can help generate revenue for their operation and maintenance and promote financial sustainability. This can include charging fees for the use of public toilets, sewage connections or wastewater/faecal sludge treatment services. Tariffs can also be used as collateral for loans or other investments.

Tariffs are an important component in full cost recovery of sanitation provision. However, only around 30% of countries applying water tariffs recover at least 80% of the operation and maintenance costs (Figure 5.7; WHO, 2022). Cost recovery rates vary greatly within countries and municipalities, this

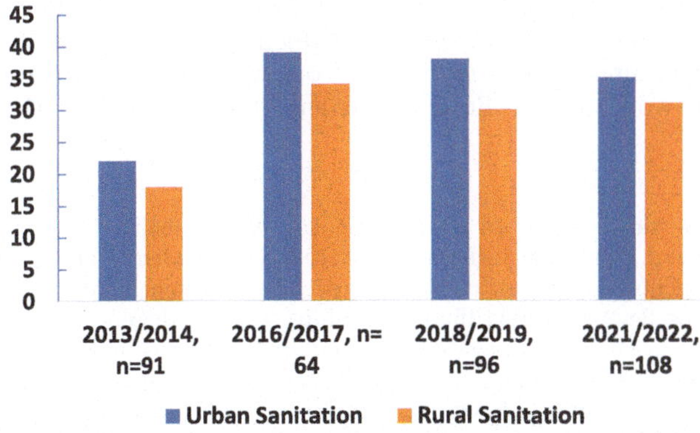

Figure 5.7 Percentage of countries indicating that more than 80% of operation and maintenance costs are covered by tariffs (*source*: with information from WHO, 2022).

being associated with the economic capacity of the region and the existence of systems to collect fees.

Applying tariffs for sanitation is a challenging task. The amount charged can be significant, for instance, in Minas Gerais, Brazil, where 42.5% of the water bill is related to the sewage tariff (Resolution No. 154/2021). However, tariffs must also be affordable. People must be willing to pay, which requires education and awareness of the effects on human and ecosystem health of water contamination from wastewater. Often this results in people believing that the costs should be borne by the government to maintain the public and social good. In contrast people paying to empty their own latrines is a logical cost; if the latrine is not emptied, the negative impacts directly affect the household. The discussion on sanitation tariffs is intertwined with privatisation, which is often rejected because, as a human right, it should be provided by governments. With this in mind, the discussion on the need to have tariff reforms[8] and independent entities to set sanitation tariffs (Wu *et al.*, 2016) is not considered in this book as first it is necessary to sort out to what extent the sanitation services are a human right and to what extent a public service. In any case, when sanitation tariffs are implemented, it is considered that a non-regulated competence (free market) for this sector is not an option and that to keep the services affordable cross-subsidies should be implemented (GWP, 2000; OECD, 2015). When sanitation tariffs are associated with wastewater reuse, it is easier to understand and accept a tariff, but by the benefitted person (which is not necessarily the same person producing the wastewater/faeces).

International aid[9]

Although donor support may be a small proportion of overall WASH funding, in around one-third of countries its support is significant (greater than 25%). Indeed, international development organisations and donor agencies provide financial support for sanitation in the form of grants, concessional loans (Box 5.2) or technical assistance. During the Water Supply and Sanitation Decade (1981–1990), investments were financed mainly by donor agencies, but the actual improvement in service coverage and the operations of water supply and sanitation facilities in the Global South remained modest. As a result, the recent focus of external support agencies has been on capacity development and policy and legal framework support (Brown & Heller, 2017; WHO, 2022). Development

---

[8] Tariff reform is considered as an essential element of sustainable financing. Wu *et al.* (2016) recommend that these tariff reforms occur independently and in advance of any PPP agreements to demonstrate the enhanced service benefits of increased tariffs without engaging fears regarding the partnerships. Indeed, the decisions around who pays for what, how much they pay and their willingness to pay represents a challenge when sanitation services are essentially a public good (Augsburg & Sainati, 2020).

[9] The use of the term 'aid' in this section is inclusive of official development assistance (ODA) grants, ODA loans and private grants, but does not include non-concessional lending. Aid is measured with the ODA in the way it was defined for the monitoring of SDG target 6.a (WHO, 2022).

> ## Box 5.2 Use of loans for sanitation in China (Kitano & Qu, 2021)
>
> To promote sustainable development of the sanitation sector, long-term foreign loans and technical cooperation are needed. An example of urban sanitation in Asia is the People's Republic of China (PRC) receiving financial assistance from Japan to build its first WWTP that was completed in 1993. The Beijing Gaobeidian Plant with a capacity of 500,000 m³/day was then the first WWTP using foreign loans (Kitano & Qu, 2021). The completion of the plant led to improved water quality in Tonghuihe River with the use of treated wastewater as cooling and irrigation water.

banks are best placed to resource the sanitation gap given their focus on financing for economic and social development that sets them apart from commercial banks (Hutchings *et al.*, 2018). However, a careful analysis needs to be made by countries to ensure that support fully fits local needs. Sometimes, development agencies when including components to provide technical assistance, capacity building and human resources education, include methodologies and points of view that are not fully aligned with national and regional objectives and local communities' needs (Figure 5.8). Less than one-third of countries receiving donor funds reported that funds were fully aligned with their national plans for the water sector. Low-income countries overwhelmingly reported less alignment with national plans than lower-middle- or upper-middle-income countries that received donor funds (WHO, 2022). Funding for the sanitation sector through development cooperation is unevenly distributed, likely to benefit dense urban informal populations to the detriment of rural, urban informal or poor

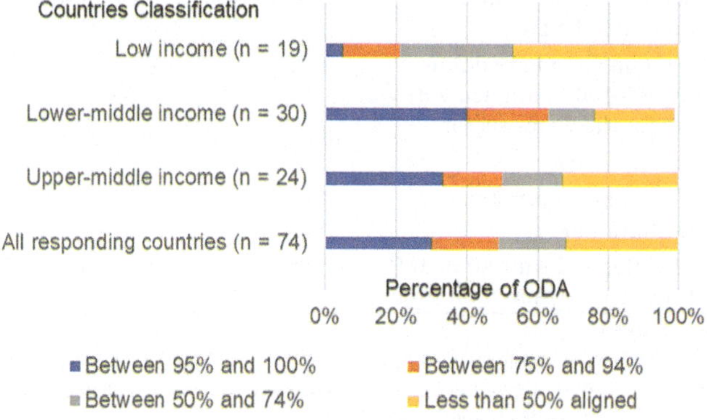

Figure 5.8 Percentage of ODA-recipient countries that reported alignment of donor funds with national water sector plans (by World Bank income group) (*source*: with data from WMO, 2022).

urban settings, allocating more funds to water than to sanitation and insufficient funds for education and training (WHO, 2022).

SDG target 6.a.1 (WHO, 2022) aims to increase international cooperation and capacity-building support for the water sector, and is monitored primarily through volume of ODA. While it covers all areas related to water, sanitation and hygiene, most of the funds (76%) are directed towards water services (WHO, 2022). This is true even though achievement of SDG target 6.2 on sanitation is lagging behind SDG target 6.1 on drinking water (Table 5.2). Between 2017 and 2020 aid for water and sanitation decreased by 5.6%, and the geographical targeting shifted from sub-Saharan Africa to eastern and south-eastern Asia (Figure 5.9). Despite this, water supply and sanitation aid provided to sub-Saharan Africa still receives the largest share globally (WHO, 2022).

Private funding

One way to respond to the lack of funding for sanitation services is the use of private companies (Tsillas, 2015). Depending on the model, revenue for private companies comes from construction or leasing infrastructure or wastewater tariffs. However, as previously noted, tariffs are not usually sufficient to cover infrastructure and maintenance costs.

PPPs and blended financing

PPPs involve collaboration between the public and private sectors to finance and manage sanitation projects. Private companies can provide funding,

Table 5.2 External support and international cooperation (SDG target 6.a).

External support and international cooperation (SDG target 6.a)	2018	2019	2020
SDG target 6.a: International cooperation and capacity building			
ODA disbursements for the water sector (constant 2020 USD)	9.6 billion	9.1 billion	8.7 billion
Percentage of countries where donor funds are fully (95–100%) aligned with national plans[a] for the water sector	–	–	30
Water and sanitation aid[b]			
Percentage of total aid commitments for water and sanitation	4.6	4.6	3.6
Breakdown of water and sanitation aid commitments between water/sanitation	65%/35%	63%/37%	60%/40%

Source: from WHO (2022).
[a] A national plan for the water sector has a broader scope than a national WASH plan and covers elements in addition to WASH relevant to SDG 6, such as water quality, water-use efficiency, water resources management and water-related ecosystems.
[b] Water and sanitation aid includes specific activities related to water supply and sanitation, as well as activities relating to water sector policy and administrative management, water resource conservation and river basin development and waste management and disposal, which are listed under the water supply and sanitation sector in the OECD-CRS aid activity database (all codes in the 140xx series).

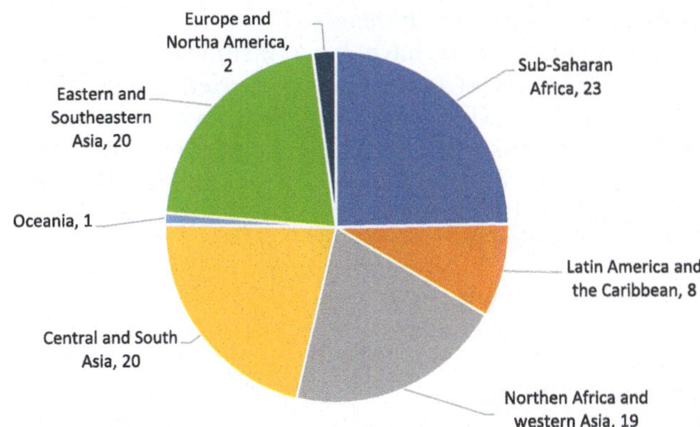

Figure 5.9 Percentage of global water and sanitation aid commitments directed to each SDG region, 2020 (*source*: with information from WHO, 2022).

expertise and technology, while the government provides regulatory oversight. PPP models can be structured in various ways, such as build–operate–transfer arrangements or performance-based contracts. PPPs can help in mobilising funds, improving service delivery efficiency and fostering innovation in sanitation financing.

In some countries, PPPs are a financing option that have a poor reputation and in many generate strongly polarised perspectives. Some of these mechanisms, particularly in the water sector, have failed to meet service objectives or to protect vulnerable populations. Furthermore, low rates of return can act as disincentives to private investment (Howard, 2021). In contrast, other nation states have managed to harness these partnerships to provide sanitation services to large proportions of populations. China leads the world in water and sanitation PPPs, accounting for 40% of these partnerships globally. This has been attributed to several factors, such as (Wu *et al.*, 2016):

- Patiently constructing appropriate commercial relationships, as private sector entities and the government work through initial poorly designed contracts.
- Strong multi-level political will and alignment of policy and legal frameworks.
- Tariff reforms that increase tariffs to 'rational' levels to ensure that the sector is commercially viable, but stopping short of full cost recovery from consumers.
- Regulations that prohibit guaranteed rates of return on investment
- Transition over time to domestic private sector investment rather than international.

Blended finance can include concessional loans, equity investments, guarantees or technical assistance from development finance institutions, commercial banks and impact investors. Blended finance mechanisms help leverage additional resources and bridge financing gaps.

5.4.1.2 Innovative financing mechanisms

Conventional economical tools have been insufficient for the needs and inadequate to fund the set of tasks and the varied groups that the entire sanitation chain implies. Thus, new options have been and are under development. These options are adapted to the participation of the different type of partners and can be at different levels (multi-level) and expand the sources for funding and their combination. These are (UN-Habitat & WHO, 2021; UN-Water, 2015b, 2024):

- National sanitation funds
- Pro-poor subsidies and grants
- Municipal bonds
- Social impact bonds
- Green bonds/funds
- Community contributions
- Microfinancing
- Crowdfunding platforms
- Philanthropy
- Corporate social responsibility (CSR)
- Innovative technology and business models
- Climate change funds
- Insurance and risk mitigation.

National sanitation funds

Establishing national sanitation funds can create a sustainable financing mechanism for sanitation programmes. These funds can be established through government contributions, donor support or innovative financing sources. They provide a dedicated pool of resources for sanitation initiatives at the national or sub-national level.

Municipal bonds

Municipalities can issue bonds[10] to raise funds for sanitation infrastructure development. The bonds can be sold to investors, and the proceeds can be used to finance projects such as WWTPs, sewage systems or solid waste management facilities. Repayment of the bonds can be structured through user fees or other revenue-generating mechanisms.

Pro-poor subsidies and grants

Subsidies[11], as explained, can be provided to reduce the financial burden on individuals or communities for constructing or upgrading sanitation facilities.

---

[10] A bond is an economic instrument that represents a loan lent by an investor to a borrower, in this case a municipality. Bonds are used by companies, municipalities, states and sovereign governments to finance projects. Owners of bond are debtholders or the issuer. As with any loan, they are subject to a deadline and interest has to be paid.
[11] Subsidies are benefits given by the government to groups or individuals, usually in the form of a cash payment or tax reduction.

Grants[12] can also be provided to organisations or institutions working in the sanitation sector to support their initiatives. The orientation towards low-income households or marginalised communities is critical to ensure that these types of funding mechanisms achieve the goal to 'leave no one behind'.

Social impact bonds

Impact bonds, also known as social or development impact bonds, involve private investors providing upfront capital to fund sanitation projects and returns are based on the achievement of predefined social outcomes. By providing capital to sanitation businesses, organisations, startups or service providers, impact investors can help scale up innovative solutions while seeking both a financial return on their investments and measurable social outcomes. If predefined outcomes, such as improved access to sanitation or reduced health risks, are achieved, the investors receive a financial return from the government or other outcome payers. Impact bonds shift the risk of investment from the public to the private sector.

Green funds

A green fund is a mutual fund[13] or another investment vehicle that will only invest in projects that are deemed socially conscious or directly promote environmental responsibility (Chen, 2022). Green funds are used to support companies engaged in environmentally supportive businesses, such as alternative water pollution control, sustainable development, sustainable buildings, human welfare and productivity increase – all topics that are related to sanitation.

Green investing began in the 1990s after the Exxon Valdez oil spill environmental disaster gained worldwide attention. With time, green funds, besides addressing environmental aspects, have started to include projects related to climate change. For green funds, profitability is not the only criteria for investors, as it includes enterprises' social responsibility, industrial image and support to develop a sustainable world.

Around 39% of green bonds issued in 2017 were for water, wastewater and solid waste management (World Bank, 2017). In 2018, a total of 100.5 billion USD were water-themed bonds. These bonds were issued mainly in Europe (63%), the Asia Pacific (19.6%) and North America (14.9%) (World Bank, 2019b).

Community contributions

Communities can contribute financially through savings groups, community-based organisations or labour contributions. This approach encourages

---

[12] A grant is an amount of money that a government or other institution provides to an individual or to an organisation for a particular purpose such as education or home improvements.

[13] A mutual fund is an investment option where money from many people, organisations or nations is pooled together to buy a variety of stocks, bonds, or other securities. A mutual fund is managed by a money manager, providing individuals with a portfolio that is structured to match the investment objectives stated in the fund's prospectus.

community involvement and ownership and ensures that the facilities meet their specific needs. Womens' groups are especially well-positioned as typically they have already set up the financial mechanisms and social enterprise required. Women also represent critical beneficiaries of improved sanitation services.

Microfinancing

Microfinance institutions (Box 5.3) can play a role in financing sanitation solutions at the household or community level. They can offer small loans to individuals or communities for constructing or upgrading sanitation facilities. Microfinance empowers individuals and communities by providing access to capital and enabling them to invest in their sanitation needs. According to the GLASS 2022 survey, women are among the primary users of microfinance solutions. For example, women represent 89% of the Water.org Water Credit programme users (WHO, 2022), evidencing that microfinancing programmes empower women.

Box 5.3 Microfinance in Ghana and Cambodia

Sanitation entrepreneurs are being encouraged to establish sanitation service businesses. The Toilet Board Coalition (www.toiletboard.org), founded in 2015, is an example of private sector investors accelerating sanitation business solutions through financial and mentorship support of small and medium enterprises. Impact stories to date include companies such as WASHKING in Ghana, which installs toilets coupled to biodigesters for households, offices and schools. Micro Enterprises for Sanitation in India establishes women and self-help groups as businesses taking care of toilet facilities.

A multi-pronged finance approach has been demonstrated to be more beneficial than sensitisation towards self-supply alone in Ghana. Results of the model indicate that subsidies for vulnerable and poor community members in association with community-led sanitation led to an additional 10,000 (25%) statistical diarrhoea cases avoided over community-led sanitation alone. However, neither option results in high net primary values, a likely consideration for prioritising investments by national government. Conditional cash transfers (demand-side subsidies) have also been effective in transforming sanitation uptake (Howard, 2021). In Cambodia, microfinance loans increased willingness to pay for latrines, even in the absence of subsidies, over those who were not given access to loans (Yishay *et al.*, 2017). Use of deposits and flexible savings through mobile money systems was found not to drive increased numbers of people investing in improved sanitation facilities in the same way that subsidies or deferred payment plans did (Lipscomb & Schechter, 2018).

Source: with information from Crocker *et al.* (2017, 2021).

Crowdfunding

Online crowdfunding platforms can be utilised to raise funds for sanitation projects, particularly for smaller-scale initiatives or community-level projects. For these platforms to be successful there is a need to raise awareness on its advantages and ways of operation for both the community receiving the support, and the civil society capable of providing it. The creation of 'crowd donors' and multi-level participation are part of the social empowerment process.

Philanthropy

Philanthropic organisations and foundations may offer grants or donations to support sanitation efforts, especially in areas with limited access to traditional financing sources. However, potential donors are not always fully aware of what is happening locally and funds go towards constructing projects rather than to ensure their proper running.

Corporate social responsibility

Companies can allocate funds from their CSR programmes to support sanitation projects for their own employees or for the local communities where they are based or extracting natural resources from. This can involve investments in sanitation infrastructure, education and awareness campaigns or partnerships with local organisations working on sanitation issues. CSR initiatives contribute to corporate sustainability goals while addressing societal needs.

Innovative technology and business models

Given the potential for reuse and resource recovery in WWTPs, the industry can develop innovative financial and business models that take advantage of these additional potential revenue sources. The most common and easy to implement scheme is the use of recycled water for agricultural irrigation or industrial use, but the range of application and regions for reuse expands with economic diversification and increased water stress. Possible areas of intervention are commercial use, urban landscape irrigation, groundwater recharge, environment and recreation, energy production and advanced treatment for potable use.

Increasingly, *closing the loop* between waste as waste versus waste as a resource has been demonstrated to provide profits that can be used to offset capital costs over time periods favourable for external investment from the household to the large urban scale (Box 5.4) For example, decentralised wastewater treatment systems can be designed to produce biogas or fertiliser as by-products, which can be sold for additional income.

At a larger scale, compost or biogas from WWTPs generate possible market value. Compost can be used as a soil conditioner or as a fertiliser. Biogas is an energy source that could be used on-site to reduce electricity demand for treatment processes. It should be noted that large urban centres have been taking advantage of the wealth in wastewater effluents for decades through anaerobic digestion, including reusing the energy generated to power wastewater treatment processes.

> ### Box 5.4 Selling biogas, fertiliser or solid fuel briquettes to create incomes for sanitation
>
> Waste to Wealth, a collaboration between various ministries in the Government of Uganda, the United Nations University Institute for Water, Environment, and Health, and Anaergia, with support from Grand Challenges Canada, established the economic, livelihood, health and environmental benefits that can accrue through use of anaerobic digestion and use or sale of the biogas, fertiliser or solid fuel briquette by-products as well as reuse of water for irrigation or groundwater recharge. The technology is not new, but reframing waste as wealth, that is, within a circular economy, provides opportunities for innovative financing mechanisms, especially where upfront capital costs are prohibitive to accessing sanitation facilities. This is especially true when many of the return-on-investment scenarios demonstrate payback within 12 months for household, school and healthcare facility systems and 5 years for larger community systems, even when costs of operation and maintenance are accounted for.
>
> *Source*: with information from Schuster-Wallace *et al.* (2017), Toilet Board Coalition (2019).

More recently, it is financially feasible to reclaim nutrients within the wastewater treatment process, with several companies generating fertiliser products from municipal wastewater treatment facilities. In a social enterprise model in smaller communities, profits can be reinvested into expanding water, sanitation and hygiene services or other social services. At a household or institutional level (school or healthcare facility), the slurry can be used as a fertiliser in kitchen gardens to increase food availability and diversity for improved nutrition. Phosphorus is a limited resource that may soon be depleted unless recycling measures are taken (SWIM-SM, 2013 - Sustainable Integrated Water Management Program). Energy and nutrient recovery could reduce the need for other forms of energy consumption (e.g. commercial fertiliser production) (Dickin *et al.*, 2020).

There are a few other innovative options to raise funds that are currently emerging. For instance, toilet designs can incorporate fee collection mechanisms or advertising opportunities. However, new business models should be considered carefully. For instance, monetising consumer data (behaviours, preferences, wastewater analytics) (Toilet Board Coalition, 2019) is an emerging revenue generating opportunity that has ethical and privacy considerations that need to be fully understood before wide-scale application.

Climate change funds

Fortunately, drinking and sanitation services are measures for adaptation to climate change in many sectors. This means that there are sources to increase sanitation by using climate change adaptation funds (Caretta *et al.*, 2022). For example, health and environment climate change adaptation strategies involve the need to increase sanitation. Further decreases in water availability as a result of climate change can partly be managed through increasing practices coupling sanitation with water reuse. Another example is the use of funds to adapt cities to an increasing flooding environment by building sustainable urban drainage and sewer systems (Caretta *et al.*, 2022).

While the amount (and percentage) of water and sanitation aid disbursement marked as climate change adaptation has ranged from around 500 million USD to 750 million USD (or 7–11%) between 2010 and 2020 (corrected to 2020 USD), the amount for adaptation had increased from approximately 600 million USD to approximately 215 million USD (i.e. from 10 to approximately 32%) for the same period of time (WHO, 2022).

Mitigation (Box 5.5) garners the significant share of committed climate finance (Caretta *et al.*, 2022). For example, of the total 15.4 billion USD for climate finance commitments through 'green bonds', 79% accrued to mitigation and the rest to adaptation (World Bank, 2017). However, within adaptation finance, water garners a significant share of funds (13% for water management, 12% for coastal management and 10% for disaster risk reduction; Adaptation Fund, 2018). Similarly, within the urban adaptation funds which receive ~3–5% of total adaptation finance flows, of 30.8 billion USD tracked in 2017–2018, water and wastewater management projects received the largest share (761 million USD annually) followed by disaster risk management (323 million USD) (Richmond *et al.*, 2021). Private financing remains a minor source of adaptation financing (World Bank, 2019a). Around 39% of green bonds issued in 2017 were for water, wastewater and solid waste management (World Bank, 2017).

So far, a greater proportion of international aids towards water and sanitation has been used for adaptation than for mitigation projects, with the aim to increase infrastructure and practice resilience (WHO, 2022). To take advantage of this kind of support it is important to select low-regret or hybrid solution sanitation approaches or technologies (section 3.3.7.2).

Insurance and risk mitigation

Insurance products can play a role in mitigating risks associated with sanitation projects, such as construction delays, cost overruns or natural disasters. Insurers can offer coverage tailored to the sanitation sector, reducing the financial burden on project developers and lenders.

5.4.2 Considerations to enable an adequate financing framework

The considerations presented below are classified into three groups: (a) the philosophy behind financing criteria; (b) the attention given to specific groups and (c) technical aspects.

5.4.2.1 The philosophy behind financing criteria

Need to develop financing mechanisms for all

The fact that there is a need to activate the entire sanitation chain in which diverse entities (big, medium, small and unipersonal enterprises, PPPs,

Box 5.5 Sources of income to provide sanitation in Mexico

Water tariffs are the main source of income for the provision of water services in Mexico. Water tariffs are set considering a fixed amount for up to 30 m³ per month – equivalent to the consumption of a four-member family – plus a charge that increases exponentially to avoid excessive water consumption. The fixed amount can be subsidised for vulnerable social groups to ensure their human rights to water. In some states, there is a third component for water tariff that is linked to sanitation. From a survey of 40 cities conducted in 2021, this tariff goes for drinking water from 0.33 USD/m³ for the 30 m³/month to up to 1.6 USD/m³. The sanitation tariff exists in less than half of the cities and goes from 0.01 USD/m³ to up to 0.75 USD/m³.

As in many countries, the investments for drinking water are higher than those for sanitation (around double). As a strategy to increase investments for sanitation, parts of the projects are financed through the National Mexican Programme to mitigate global warming. GHG emission is mitigated through: (a) wastewater treatment; (b) capturing the methane produced during wastewater and sludge treatment and producing energy with it; (c) using solar energy generated on-site; (d) giving priority to projects implying a lower energy demand; (e) installing low-energy consumption equipment (notably pumps and turbo blowers) in WWTPs. In 2021, through these activities, a release of an amount of 12.04 million tonnes of equivalent of carbon dioxide was mitigated.

Energy consumption in WWTPs represents up to 70% of the operation costs. This high cost is the main reason several WWTPs have stopped operating. Therefore, cogeneration and use of solar energy are part of the national strategy to increase, foster and sustain the sanitation coverage in the country. Cogeneration is achieved using the methane produced in WWTPs, as mentioned before. In 2022, the total amount of energy generated in this way was amounted to 26 MW. Solar photovoltaic cells are used in small WWTPs, amounting to a total production of 3.1 kW, saving from 15 to up to 100% of the total energy consumption.

CONAGUA runs programmes with the specific aim to address the needs of the most vulnerable populations. Thanks to this effort, in 2021, 23,944 inhabitants were connected to sewer systems and the sewerage for 48,889 inhabitants was improved. In addition, 3,041 dry toilets and biodigesters were installed.

To improve sanitation services, the communities can request the government to receive back the payments made for the disposal of used water (principle of who pollutes pays). This way in 2021, 56 million USD were allocated additionally for sanitation.

Source: with information from CONAGUA (2022).

academic and research centres, NGOs, social enterprises, community groups, the households themselves) with different forms of organisation perform a wide variety of activities implies the need to develop and implement a wide variety of funding mechanisms. There is a need for governments to ensure this.

As part of the financing mechanisms, it is important to promote the adequate collection of municipal taxes (any kind) that in the longer term will generate a viable revenue stream to refinance sanitation works, rendering sanitation utilities eligible for loans and capable to use debt instruments notably for future investments in new, more expensive infrastructure.

As the self-supplied group represents a significant proportion of sanitation financing, special attention must be paid to their financing mechanisms. This has been demonstrated in India, where 80% of new toilets were self-funded, with informal loans (supporting 9% of toilet acquisitions) and government subsidies (supporting 8%) (Augsburg & Rodríguez-Lesmes, 2020). The study further recognised the importance of social marketing (e.g. no latrine, no bride) on decisions to invest in sanitation, with the addition of a woman to a household or the presence of a man close to the legal age of marriage increasing likelihood of investment in a toilet as a result (Augsburg & Rodríguez-Lesmes, 2020).

Transparency and accountability

It is crucial to ensure that funds are allocated appropriately, financial mechanisms are transparent and accountable and investments are targeted towards areas with the greatest need to achieve universal access to safe sanitation. For this, the government must develop and implement proper transparent mechanisms for both the public and private sectors. This way it is possible to oversee whether funds have been appropriately used by public and privately managed sanitation utilities. If necessary, instruments to allocate civil and penal responsibilities must be available and accessible. The methods set for tracking the origin and end of funds are also useful to clearly understand who is paying for what and to set equity criteria for the allocation of public funds.

5.4.2.2 Critical groups for financing

Women

Women represent the majority of the poor and, in patriarchal societies in particular, may not have decision-making authority in the household or access to funds to be able to purchase sanitation (WHO, 2022). Community sanitation enterprises, particularly through women's groups, can be powerful facilitators for access to sanitation. In addition to valuable by-products associated with waste, hygiene products such as soap can augment income for individuals and community groups. Women's groups represent particularly beneficial recipients of financial and capacity investments towards sustainable and universal sanitation. Existing systems such as rotating savings clubs and enterprising activities that women's groups undertake mean that the financial mechanisms and social enterprise required to support local sanitation self-supply and pro-poor supply already exist in many instances.

Poor people

Although many nation states include pro-poor strategies in development plans or sanitation strategies, the reality is that these mechanisms do not always reach the poorest of the poor on the ground (e.g. Bisung *et al.*, 2016). Pro-poor mechanisms include those directed at the individual household (reduced tariffs, special financing mechanisms), institutional arrangements to provide broad sanitation services (e.g. to urban poor or an informal settlement) and design of pro-poor regulatory and monitoring systems to ensure that programmes actually serve the poor (Hutchings *et al.*, 2018). Although 80% of all countries have specific measures in policies and plans to reach people living in poverty, just over half have corresponding measures for monitoring and fewer have finance measures that are consistently applied (WHO, 2022).

5.4.2.3 *Technical aspects*

Selection mechanisms

For the sustainable management of sanitation, it is necessary to choose not only technically, economically and financially viable options (UN-Habitat & WMO, 2021) but also solutions that are socially well acceptable even for those that are not directly benefited from the project. Many sanitation projects have been stopped for social reasons, which are collectively described as the syndrome of 'not in my back yard'[14].

Mixing financing mechanisms and blending funding sources

Certainly, for sanitation, a single mechanism is not the best option and the use of multiple financing mechanisms provides the flexibility desired to address national needs (Boxes 5.6 and 5.7) and local circumstances. Having a significant portfolio of options that municipalities and communities can tailor to their specific evolving circumstances will be an asset for governments to provide sanitation for all. However, it is important to note that the choice and combination of financing mechanisms and sources for funds will depend on the local context, available resources, cultural background, affordability and the specific sanitation challenges to be addressed. Efficient and sustainable sanitation can be achieved using a multi-partner approach and, when and where relevant, strong partnerships between government, private sector actors, development partners and local communities for implementing effective and sustainable financing mechanisms for sanitation are in place. The use of mixed financing mechanisms entails the need to ensure all are funded and well-coordinated.

---

[14] An expression signifying your own or somebody else's opposition to the locating of a facility in one's neighbourhood. The phrase seems to have appeared first in the mid-1970s, when towns opposed the location of nuclear power plants and other big development projects near to them. It has also been applied to water reuse and wastewater treatment projects, even if local people support pollution control and the efficient use of water.

Box 5.6 Evolution of the financing mechanisms for sanitation in Brazil

The history of advances in the sanitation structure began in the 1960s, with the creation of the National Housing Bank (BNH), which made significant investments in the sector. Over the last 60 years, much has been done in terms of investments to support universal access to water supply and sewage services, but there is a long way to go to meet overall needs, especially considering Brazil's continental characteristics.

According to Borjas (2014) the financing of public basic sanitation services in Brazil has been enabled by different sources and forms of resource allocation, such as grants, subsidies with national budgetary resources, direct investments of public and private capital, loans from public and private funds, agencies' multi-lateral agreements (tax exemptions and taxes on services, among others). The resources come from the Fundo de Garantia do Tempo de Servicio (FGTS), the Worker Support Fund and multi-lateral agencies, such as the World Bank (IBRD), the Inter-American Development Bank (IDB) and the Japanese Cooperation Bank (Table B.5.1).

Table B.5.1 Main financing sources of water and sanitation in Brazil (*source*: Borjas, 2014).

Type	Source
Non-costly resources	General budget, public grants, treasure (union, states, municipalities and federal districts).
Costly resources	Funds managed by the Federal Government (FGTS and FAT/BNDES which are Brazilian fiduciary funds).
Service provider resources	Taxes and tariffs.
National system resources of water resources	Charge for the use of water resources.
Outside loans	Loans from international organisations (IDB or Inter-American Development Bank, IBRDI or International Bank for Reconstruction and Development, JBIC or Japan Bank for International Cooperation, KfW or Kreditanstalt für Wiederaufbau).
Private resources/ instrument	Partnership with the private sector. Real-estate entrepreneurs. Debentures. Stocks and bonds. Credit Right Funds (FIDC). Real-Estate Investment Funds (FII).

Regarding investments, two programmes need to be highlighted. First, the *Sanitation for All* (Saneamento Para Todos, in Portuguese) established in 2005 and managed by CEF (Caiza Economica Federal, in Portuguese), financing sanitation projects with FGTS resources, during the BNH era. The main ongoing programme for financing the sanitation sector is the Growth Acceleration Programme (Programa de Aceleração de Crescimento (PAC), in Portuguese), established in 2007, encompassing several infrastructure sectors, such as logistics, energy and urban infrastructure, which includes, among other areas, housing and basic sanitation. According to the 7th PAC report (2015–2018), 50.3 billion reais have already been invested in sanitation works, covering 3,753 municipalities (Correia *et al.*, 2020).

Article 8 of the new Decree no. 10.710 details the plan for fundraising resources. The service providers must indicate the financial agents and the strategies to finance their investment projects. The financing structure in Brazil is aligned with the characteristics of the sector's needs and difficulties in obtaining resources for investment, which lead to the prioritisation of the use of internal sources to finance projects, as can be seen by the company's low leverage. In this context, it can be concluded that collecting resources through tariffs plays a crucial hole in the expansion of sanitation services to achieve the universality established in the new legal framework. According to Cicogna *et al.* (2022) 'this is a controversial point for expanding the sanitation infrastructure in regions further away from urban centres. Investments with very long terms, added to the higher cost of capital in this type of funding, can lead to the need to increase tariffs, making it unfeasible to offer services in regions with lower per capita income, even if there is socialisation of costs to the entire region served by one provider'.

It is up to regulatory agencies to define tariff adjustments periodically, with the aim of mitigating risk, stabilising the sector and, consequently, attracting new investments. Brazil has almost a hundred entities that regulate sanitation services with municipal, intermunicipal, district or state operations. These institutions separately or jointly regulate basic sanitation services: water supply, sewage collection and treatment, urban solid waste management and urban rainwater drainage and management (ANA, 2024). The tariff is only charged in areas where the service is available, as established in Federal Law 11445. In general, the sewage tariff costs the consumer around 80% of the water tariff, a percentage recommended by the Brazilian Association of Technical Standards, as a return coefficient, as 20% is lost in watering gardens, evaporation, food consumption, among others.

Source: with information from Borjas (2014), Santos *et al.* (2018).

Box 5.7 Options for financing the Tunisia Water Reuse Master Plan 2050

The costs of the action plan to reuse water in Tunisia have been assessed and estimated over 30 years at 3.8 billion USD, including 0.96 billion USD for initial investment, 1.9 billion USD for renewal (64 million USD/year) and 0.9 billion USD for operation (32 million USD/year).

The funding system was selected based on meeting the following four objectives: (1) a financial objective: to achieve recovery of the full costs of the service to ensure the sustainability of the service in the medium term; (2) a social objective: the tariff system put in place should be accepted by the users; (3) an environmental objective: the financing system must aim at better management of water demand in order to take into account the scarcity of the resource and encourage water savings and (4) an economic objective aiming at the efficient allocation of water resources. The different options for financing the Water Reuse Master Plan 2050 that have been identified are:

- Financing through reclaimed water pricing could recover almost 20% of total costs over 30 years.
- Financing through the implementation of a reuse fee for sanitation users at the national level so that all water users participate in the financing of the water reuse. A 20% increase in the sanitation fee (from 0.22 to 0.26 USD/m^3) would finance 10% of the reuse cost over 30 years. The total amount paid by users on the water bill (drinking water + sanitation) is currently around 0.48 USD/m^3 on average, all users combined.
- Financing via an environmental tax, a 'water recovery tax', to be applied on each tourist night, mirroring the already existing tourist tax. The development of water reuse will undeniably improve the quality of bathing water and more generally the coastal environment. Tourists will in practice be beneficiaries of the country's water reuse policy. This tax would finance between 4 and 23% of the 30-year cost of the water reuse.
- Funding via donors and/or the Green Climate Fund.
- Financing through PPPs.
- Financing via local taxes or a national tax.

Data and monitoring

In section 5.4.1.1, it has been argued that public funding is indispensable to leave no one behind. But, to be able to direct the efforts and verify that this is happening as planned there is a need to have adequate monitoring and reporting systems. Ensuring sanitation for all requires countries to identify

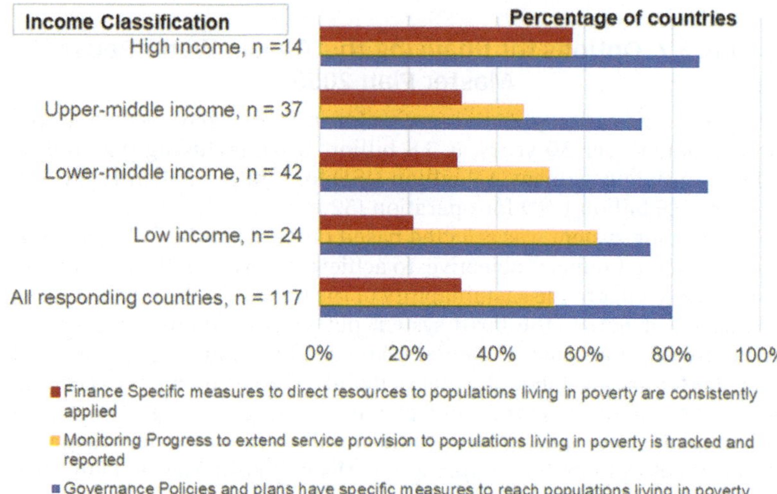

Figure 5.10 Percentage of countries with measures in policies and plans, which monitor service provision and direct financial resources to improve and extend sanitation services to specific populations and settings (*source*: with information from WHO, 2022).

and target measures of populations and settings that are being left behind, which vary from country to country. Figures 5.10 and 5.11 show that not many countries have set measures in policies or plans to address vulnerable population however, and even fewer monitor or report the progress when available (WHO, 2022).

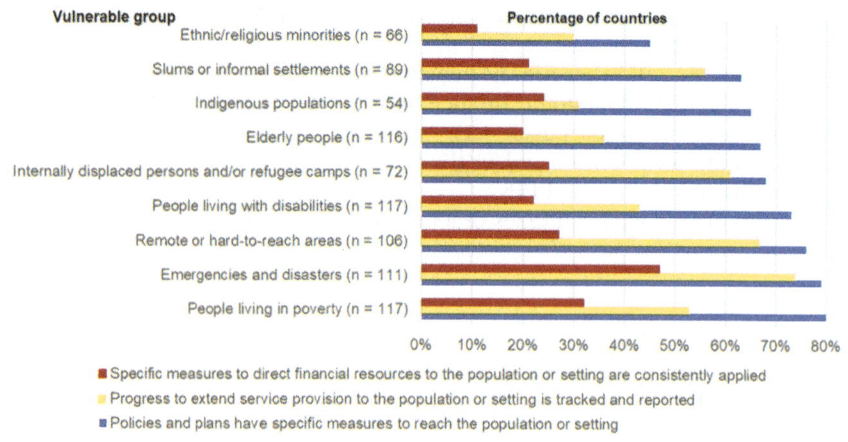

Figure 5.11 Measures to extend sanitation services to people living in poverty (*source*: GLASS, 2022).

5.4.3 Strategies for disbursement

The three strategies for disbursement are not mutually exclusive:

- Conventional disbursement
- Revolving funds
- Results-based financing (RBF)

5.4.3.1 Conventional disbursement

In this case, a budget is presented, funds are allocated and periodically the organisation disburses money to the entity for usage. Funds are usually allocated periodically throughout the year according to a calendar agreed in advanced. If an organisation does not manage to spend all the funds according to the calendar in many cases the budget is simply cut, regardless of the sanitation needs.

5.4.3.2 Revolving funds

Revolving funds are financial mechanisms that use repaid funds to finance new sanitation projects. For instance, a community-based organisation can establish a revolving fund where community members can borrow money for constructing toilets. As loans are repaid, the funds are recycled and made available to other community members.

5.4.3.3 Results-based financing

RBF mechanisms tie financial disbursements to the achievement of predefined results or outcomes. For example, funds may be released based on the number of toilets constructed, the percentage of households with access to improved sanitation or the reduction in waterborne diseases. RBF incentivises performance and ensures accountability.

5.5 EXPENDING FUNDS

Many of us have heard the expression 'if you think it was hard to get the funding wait to see how hard it is to expend it...' and unfortunately this is true. As with many government activities, the provision of good services depends not only on the effective planning and how funds are raised but also on how efficiently they are implemented (UN-Water, 2024). To ensure the speedy and appropriate use of funds it is necessary to include administrators and public inspectors who understand the challenges to provide sanitation for all, notably to vulnerable groups. They also have to have an understanding of acquiring additional funds from the diversification of 'sanitation utilities mandate' in order to reuse water and reclaimed by-products. This will prevent administrative procedures becoming a bottle neck when funding is available. Even when an activity is budgeted and effectively funded (*WASH budget*) this does not mean that the government spends the funding (GLASS, 2022) *WASH expenditure* is the money that is actually spent on WASH by governments, external sources, households and repayable finance (GLASS, 2022). Thus, even if WASH budgets may be increasing, governments may be limited in their spending by how well budget allocations can be absorbed by the relevant ministries. Less than half of countries where acceleration in coverage is needed reported over 75% absorption of domestic commitments for all four subsectors. Half of the countries reported using less than 75% of domestic capital commitments for urban and rural drinking-water supply and sanitation. Lengthy and complex procurement processes were most often cited as obstacles in improving the efficient and timely use of domestic capital commitments for WASH (WHO, 2022).

As presented, financing and procurement are not easy tasks and are highly technical. Policy and decision makers must have a good grasp on the subject and as much support as possible from administrators sensitive to sanitation and to the challenges that reaching 'all' represents.

doi: 10.2166/9781789064049_0143

Chapter 6
Conclusions

Current sanitation situation and its causes: Currently, only 25% of countries are on track to achieve their sanitation targets (WHO, 2022). This 'sanitation crisis', as some like to call it, was first attributed to a lack of funds, then to poor 'governance', later to the need to have private participation and then to poor management of private and public participation. Since then, the list of causes kept increasing. Maybe it is time to think that the sanitation crisis observed in the Global South is due to its complexity and clear association with poverty and inequity. Simply, if all people were wealthy enough, they themselves – with nobody's advice, assistance and even completely ignoring the causes of their situation – would arrange 'first world class' sanitation services for themselves.

In addition to the financial implications, sanitation services are not a 'politically sexy' topic, nor a 'good subject' to discuss in public. Furthermore, those with the public well-being in their hands do not advocate or support it. For many women, and also to some men, who have been in charge of daily changing their children's nappies this is hard to understand.

In many high-income countries, where sanitation has been provided successfully for several decades, governments have had plenty of funds and have done it through administrations that have mostly not been restricted by the very long list of recommendations that the Global South keeps receiving to perform the same task; for example, reducing greenhouse gas emissions and minimising water pollution. And, looking at the challenge, and at the list of recommendations to be addressed, it is clear that fulfilling sanitation for all will be a medium- and long-term task.

Fortunately, considering the importance that many countries place on achieving the 2030 agenda, sustainable development goal 6 and target 6.2, sanitation is improving. The success is now to be evaluated not only as the increase in coverage, but also on the achievement of a good quality and sustainable service. This is particularly true as many of the facilities once installed have stopped functioning due to the lack of access to affordable energy, lack of water to run the facilities, poor design and construction and the unsuitability of designs not adapted to cultural and social conditions. Simply put, sanitation facilities for the poor need to be as comfortable and dignified as those available for the rich. Otherwise, services will be understandably rejected and there will be no sanitation.

For all: Worldwide, the most challenging aspect of providing sanitation is the aim to 'reach all'. This requires tasks which were not commonly performed, or the appropriate information was unavailable. Identifying and locating vulnerable groups, understanding their needs and finding options to address them demands a careful analysis and planning, funds availability, capable human resources and efficient implementation.

General management: Under the conditions that prevail in the Global South (or Majority World, to use a more recent term) sanitation-related authorities (from any sector) will need to take the best possible and timely decisions using whatever information is available. In addition, they will need to collaborate with interested parties from all sectors and with the actors of the entire sanitation chain to implement actions adapted to the local sociocultural conditions and realistically considering technical and financial capabilities. Furthermore, for non-stable societies, they will need to consider the political situation. Policy and decision makers need to master tools for cooperation, as collaboration along the entire sanitation chain is key. This means they need to be team players and work with different sectors and across government levels, users and people affected or benefitted by their projects. For this, they must have good knowledge and understanding of the role each institution and stakeholder formally and politically undertakes.

Administrative and legal frameworks: Developing a portfolio of differentiated solutions to address urban and rural needs is key. For this portfolio to be applicable, regulations and administrative procedures must be flexible by considering new trends in sanitation. For instance, the concept of circular economy (still under evolution) which makes sense, especially to Indigenous communities for whom living in harmony with their environment has always been part of their traditions. Circular economy also makes sense considering that water is a resource that is not destroyed when used, and only needs to be cleaned for reuse, like a dress. But, to undertake water reuse programmes, such as for agricultural irrigation, the participation of other sectors and stakeholders is essential to ensure health protection, agronomic needs, exportation requirements and farmers' involvement. Bringing all these participants into new practices demands communication, education and training. This includes the beneficial reclamation of the compounds that are contained in wastewater or produced during its treatment.

The sanitation chain comprises a series of services that are placed downstream of the user. Many sectors, administrative levels and a very diverse set of stakeholders and partners participate in it. The legal framework can be very useful to clarify mandates, set coordination mechanisms and, most of all, to provide the local governments, those who implement sanitation in the field and have the direct responsibility towards the users, a proper framework to perform their duties. To avoid a mismatch between the legal framework and the current practices in many countries of the Global South, compliance needs to be enforced.

Stakeholders' and partners' involvement, participation and coordination: The main objective for the involvement of the public is to ensure that everybody is aware of the role they and others play, including their rights and responsibilities. The government must retain the coordination of the processes respecting the accountability principle. For the process to be effective, policy and decision makers must gain the confidence of sanitation users and stakeholders by keeping interested parties well informed, maintaining individual motivation, demonstrating organisational commitment, promoting public communication and dialogue and ensuring a fair and robust decision-making process and outcomes. It is also important to act diplomatically, because not only what is being said matters but also how it is being said. Water policy and decision makers need to be the first to voice that sanitation brings well fare to people and communities in areas other than water, such as health, education, gender equality, food security and economic development.

In the Global South, it is important to keep in mind that when the public's involvement in sanitation projects is limited it does not mean that information should not be given or their local sociocultural needs disregarded. Sanitation is a human right and its provision should be provided equally and in a timely manner for all communities, regardless of their sociocultural-economic standing, not making differences in the scheduling of task nor in the quality of services for any community. In many places of the Global South, both men and women – notably in regions where mono-parental families prevail – have long commutes to meet their minimal economic requirements and when time is available, they need to choose, among many other activities, public participation, including changing the ruling political party, performing personal administrative procedures or chores, spending time with family or participating in religious events. Under such circumstances, it would be wise to take advantage of these other activities to raise awareness on sanitation, and eventually to encourage active public involvement in its design and implementation.

Financing: Conventional mechanisms for financing have been insufficient so far to fund sanitation. Fortunately, non-conventional financial mechanisms have been developed, some of them still under evolution, and are adapted to cover financial needs of all the components of the sanitation chain.

Gender approach: For sanitation, the role of women is especially important due to their leadership in communities and their ability to organise and guarantee the sustainability of systems. But as well, their participation in the decision-making process is essential. Women's caregiving role at home and as part of a community has made them understand that sanitation is relevant, they often do not fear talking about it publicly and are ready to raise it to a high level in the political agenda.

Ultimately, we need to rethink sanitation in terms of who has access, what levels of access, sustainability of access, and protection of human health and the environment. As this book has outlined, this rethinking requires different people, different solutions, different financing mechanisms, and different supporting legal and institutional frameworks. Together, we can make sure that no-one is left behind.

Definitions

Acceptable sanitation facilities (UN definition) Any kind of improved sanitation facility can potentially be safely managed, but only septic tanks and sewer lines are included in the definition of safely treated wastewater. This is because all households generate wastewater, including blackwater (from defecation and urination) as well as greywater (from other domestic uses, including washing and bathing). Safely managed sanitation is concerned with safe management of blackwater, but safely treated wastewater considers both blackwater and greywater. Sewer lines and septic tanks, unlike pit latrines, have the potential to manage greywater as well as blackwater flows. In principle, greywater could also be safely treated separately from blackwater (for example, through household or community soak pits).

Acceptable treatment (UN definition) Secondary treatment processes or higher are adequate for safely managed sanitation services and are sometimes also used for calculation of safely treated wastewater. However, additional data on compliance of treated wastewater with relevant limits (for example, effluent quality standards) are used for the SDG indicator 6.3.1 when available.

Accountability Relationship among actors that has five features: delegation, finance, performance, information about performance, and enforceability. Public accountability is between the government and society, and commercial accountability refers to the client-users relationship (World Bank, 2003).

Adaptation Climate change adaptation is defined as an activity that intends to reduce the vulnerability of resilience, through increased ability to adapt to, or absorb, climate change stresses, shocks and variability and/or by helping reduce exposure to them (WHO, 2022a).

Aid (Financial) All official development assistance. It comprises grants, loans and private grants received by a country. It does not include non-concessional lending. It is measured with the ODA (Overseas Development Aid).

B

Basic sanitation (UN definition) Use of improved facilities that are not shared with other households.

Basic sanitation (this book) In this book basic sanitation is understood in the way WHO (2018) defines sanitation to avoid the discussion of the complex WHO-UNICEF JMP Sanitation ladder: *"the access to and use of facilities and services for the safe disposal of human urine and faeces. A safe sanitation system is a system designed and used to separate human excreta from human contact at all steps of the sanitation service chain from toilet capture and containment through emptying, transport, treatment (in-situ or off-site) and final disposal or end use"*.

Bond Loan made by an investor to a borrower. Bonds are used by companies, municipalities, states, and sovereign governments to finance projects.

Build-Operate-Transfer (BOT) Contract performed between a public and a private entity which remains in charge of the finance, design, construction, management of facilities, to operate them through a concession scheme. The private entity has the right to operate the facility for a specific period of time, during which its investment and operating and maintenance expenses are recovered through the project. These contracts are common for large-scale projects.

C

Circular Economy An economic system where materials never become waste, are regenerated to be reclaimed and eventually can become an income source. In the context of sanitation, it can apply to the entire sanitation chain, considering the reuse of water and the reclamation and recycling of byproducts.

Co-benefits Positive impacts arising from the measures taken in a single sector that benefit other sectors synergistically.

Compact The broad, long-term relationship of accountability connecting policymakers to organizational providers. This is usually not as specific or legally enforceable as a contract, but an explicit, verifiable contract can be one form of a compact (World Bank, 2003).

Complex problems A set of problems for which the cause-effect relationship is ambiguous, uncertain, non-lineal and the problem itself is permanently evolving as it receives internal and external feedback. Due to this, there is an ample set of solutions for complex problems that have different degrees of efficiency and evolve as the problem itself evolves. For complex problems, the idea is not to find 'the solution' but solutions producing stable situations for as long as possible.

D

Decentralisation The term is used to cover a broad range of transfers of the 'locus of decision making' from central governments to regional, municipal or local governments (Sayer et al., 2004).

Deconcentration   The process by which the agents of central government control are relocated and geographically dispersed (Sayer et al., 2004).

Design-Build-Operate (DBO)   It is project delivery model contract in which a single contractor is appointed to design and build a project and then to operate it for a period of time.

E

Ecological sanitation or 'ecosan'   A sanitation option that involves the reuse of human waste as a resource, rather than simply disposing of it. It recognizes that human waste contains valuable nutrients and organic matter that can be recycled as fertilizer and soil amendment, rather than being treated as a waste product.

Environmental perception   Information processing systems in which individuals actively explore their surroundings and extract and use information in constant interaction between themselves and their environment. Public perception is closely related to factors such as social, health, economic and environmental aspects that determine the population's acceptance or rejection of sanitation and water reuse systems.

Estimated costs   The cost predicted by the government to implement a plan or strategy (WHO 2022a).

Expenditure   The money that is spent from a budget.

F

Fragile contexts (or unstable contexts)   Fragile contexts are those in which the combination of exposure to risk and insufficient coping capacity of the state systems and/or communities to manage, absorb or mitigate those risks prevail. Fragility can lead to violence, poverty, inequality, displacement, and environmental and political degradation. Countries that are classified as such are Afghanistan, Angola, Bangladesh, Benin, Burkina Faso, Burundi, Cambodia, Cameroon, Central African Republic, Chad, Comoros, Congo, Côte d'Ivoire, Democratic People's Republic of Korea, Democratic Republic of the Congo, Djibouti, Equatorial Guinea, Eritrea, Eswatini, Ethiopia, Gambia, Guatemala, Guinea, Guinea-Bissau, Haiti, Honduras, Iran (Islamic Republic of), Iraq, Kenya, Lao People's Democratic Republic (or Laos), Lesotho, Liberia, Libya, Madagascar, Mali, Mauritania, Mozambique, Myanmar, Nicaragua, Niger, Nigeria, Pakistan, Papua New Guinea, Sierra Leone, Solomon Islands, Somalia, South Sudan, State of Palestine, Sudan, Syrian Arab Republic, Tajikistan, United Republic of Tanzania, Timor-Leste, Togo, Turkmenistan, Uganda, Venezuela (Bolivarian Republic of), Yemen, Zambia, Zimbabwe (OECD, 2020; JMP, 2023).

G

Government budget   These are funds that have been allocated to be spent on a specific activity such as sanitation. The budget can be aligned to plans and strategies. The availability of budget implies that the total amount must be expended.

Government of sanitation   The combination of structure, procedures and rules that are designed to manage sanitation. It has four components: the policy (with the overall vision), the institutional framework (with administrative structure and procedures), the legal framework (with the rules for the functioning of the entire government structure) and the stakeholders (social component).

Grant   Amount of money that a government or other institution gives to an individual or to an organization for a particular purpose such as building latrines.

Guerrilla   A set of persons forming an unofficial army. It usually fights against an official army, police force or government.

H

Health   Includes physical, mental and social well-being as they affect and are affected by WASH options and conditions (JMP, 2023).

Household decision-making   Individuals' opportunities to influence and make decisions about water, sanitation and hygiene within their homes (JMP, 2023).

I

Institutional framework   The set of formal organizations a government employs to regulate, provide and oversee an activity. In the field of sanitation the institutional framework comprises several organizations from the national, regional and local level. They do not all belong to the water sector.

L

Landlocked developing countries (LLDCs)   Countries with a lack of territorial access to the sea. They are Afghanistan, Armenia, Azerbaijan, Bhutan, Bolivia (Plurinational State of), Botswana, Burkina Faso, Burundi, Central African Republic, Chad, Eswatini, Ethiopia, Kazakhstan, Kyrgyzstan, Lao People's Democratic Republic (or Laos), Lesotho, Malawi, Mali, Mongolia, Nepal, Niger, North Sudan, Tajikistan, Turkmenistan, Uganda, Uzbekistan, Zambia, Zimbabwe (JMP, 2023).

Least developed countries (LDCs)   Low-income countries confronting severe structural impediments to sustainable development. They include Afghanistan, Angola, Bangladesh, Benin, Bhutan, Burkina Faso, Burundi, Cambodia, Central African Republic, Chad, Comoros, Democratic Republic of the Congo, Djibouti, Eritrea, Ethiopia, Gambia, Guinea, Guinea-Bissau, Haiti, Kiribati, Lao People's Democratic Republic (or Laos), Lesotho, Liberia, Madagascar, Malawi, Mali, Mauritania, Mozambique, Myanmar, Nepal, Niger, Rwanda, Sao Tome and Principe, Senegal, Sierra Leone, Solomon

Islands, Somalia, South Sudan, Sudan, Timor-Leste, Togo, Tuvalu, Uganda, United Republic of Tanzania, Yemen, Zambia (JMP, 2023).

Limited sanitation (UN definition)   Use of improved facilities that are shared with other households.

M

Mitigation   Climate change mitigation is defined as an activity that contributes to the objective of stabilization of greenhouse gas concentrations in the atmosphere at a level that would prevent dangerous anthropogenic interference with the climate system by promoting efforts to reduce or limit greenhouse gas emissions or to enhance greenhouse gas sequestration (WHO, 2022a).

Mutual fund Investment option where money from many people, organizations, nations is pooled together to buy a variety of stocks, bonds, or other securities.

N

Non-sewered systems (NSS)   A sanitation system that is not connected to a networked sewer system.

O

On-site Sanitation Systems (OSS)   A sanitation system in which excreta and wastewater are collected, stored and/or treated on the plot where they are generated.

Open defecation (UN definition)   Disposal of human faeces in fields, forests, bushes, open bodies of water, beaches or other open places, or with solid waste.

P

Performance-Based Contracts (PBCs) Contracts that are conceived on having results, achievements and with payments conditioned to output achievement.

Polycentric governance The concept that the government needs to have multiple centers of decision-making, or multiple authorities, none of which has ultimate authority for making all collective decisions.

Privacy An individual's ability to feel free from observation or being heard or disturbed by others when accessing and using sanitation locations and water sources, including for hygiene (e.g. menstruation, bathing) purposes (JMP, 2023).

Public participation The intervention of citizens in decision-making regarding the management of resources and actions that have an impact on the development of their communities.

S

Safely managed sanitation (UN Definition)   Use of improved facilities that are not shared with other households and where excreta is safely disposed of in-situ or removed and treated off-site.

Sanitation   The access to and use of facilities and services for the disposal of human urine and faeces. When sanitation is safe the system is designed and used to separate human excreta from human contact at all steps of the sanitation service chain from toilet capture and containment through emptying, transport, treatment (in-situ or off-site) and final disposal or end use.

Sanitation chain   The series of facilities and services needed to provide sanitation. It comprises the toilet facilities, the excreta and wastewater collection methods and processes, the transportation and treatment of water and wastes, and the water reuse or disposal.

Septage   The combination of scum, sludge, and liquid that accumulates in septic tanks.

Shared sanitation facilities (UN definition)   Sanitation facilities that are not considered part of the safely managed services options because they have limited accessibility, lack of privacy and have health impacts because they are shared.

Small island developing states (SIDS)   American Samoa, Anguilla, Antigua and Barbuda, Aruba, Bahamas, Barbados, Belize, Bonaire, Sint Eustatius and Saba, British Virgin Islands, Cabo Verde, Comoros, Cook Islands, Cuba, Curaçao, Dominica, Dominican Republic, Fiji, French Polynesia, Grenada, Guam, Guinea-Bissau, Guyana, Haiti, Jamaica, Kiribati, Maldives, Marshall Islands, Mauritius, Micronesia (Federated States of), Montserrat, Nauru, New Caledonia, Niue, Northern Mariana Islands, Palau, Papua New Guinea, Puerto Rico, Saint Kitts and Nevis, Saint Lucia, Saint Vincent and the Grenadines, Samoa, Sao Tome and Principe, Seychelles, Singapore, Sint Maarten (Dutch part), Solomon Islands, Suriname, Timor-Leste, Tonga, Trinidad and Tobago, Tuvalu, United States Virgin Islands, Vanuatu (JMP, 2023).

Smart sanitation   Sanitation solutions promoting sustainable and efficient water management, resource recovery, and improving public health in the context of smart cities.

Subsidy   A benefit that groups, organisation or individuals receive in the form of a tax reduction or cash payment from the government who is looking to motivate them to perform a specific behaviour.

Sustainable sanitation systems   Those systems that, besides protecting human health and the environment, are also economically viable, socially acceptable, and institutionally applicable (SuSanA, 2008).

T

Trachoma   Infectious disease caused by a Chlamydia bacterium, causing roughening of the inner surface of the eyelids, causing pain and eventually blindness which can be permanent.

U

Unimproved sanitation (UN definition)   Use of pit latrines without a slab or platform, hanging latrines or bucket latrines.

V

Vector   A living organism that transmits an infectious agent from an infected animal to a human or another animal. Some examples are mosquitoes, ticks, flies, fleas and lice (EFSA, 2024).

Voice (of society)   The avenue connecting citizens and politicians and 'comprises many formal and informal processes, including voting and electoral politics, lobbying and propaganda, patronage and clientelism, media activities, access to information, and so on'. Citizens delegate the functions to politicians serving their interests, politicians perform by providing services, such as law and order to communities.

Z

Zero Waste   Activities in which nothing is discarded but instead is reused or recycled. In sanitation it includes water and by products.

References

Abedin M. A., Collins A. E., Habiba U. and Shaw R. (2019). Climate change, water scarcity, and health adaptation in southwestern coastal Bangladesh. *International Journal of Disaster Risk Science*, **10**(1), 28–42, https://doi.org/10.1007/s13753-018-0211-8

Adaptation Fund (2018). Briefing Note. Adaptation-Fund-Briefing-Note-April-2018-web. pdf, www.adaptation-fund.org/wp-content/uploads/2018/04/Adaptation-Fund-Brie fing-Note-April-2018-web.pdf, accessed in February 2024.

ADB (2021a). Case Stories on Inclusive Sanitation. Changing the Future of Sanitation through Reinvented Toilets. chrome-extension://efaidnbmnnnibpcajpcglclefindmkaj/; https://www.adb.org/sites/default/files/publication/696031/case-story-4-sanitation-reinvented-toilets.pdf, accessed in December 2023.

ADB (2021b). Citywide Inclusive Sanitation Guidance Notes: Addressing Gender Equality and Social Inclusion in Urban Sanitation Projects. https://www.adb.org/publications/gender-equality-social-inclusion-urban-sanitation-projects, accessed in November 2023.

AfChoudry E. and Islam S. (2015). Nature of transboundary water conflicts: issues of complexity and the enabling conditions for negotiated cooperation. *Journal of Contemporary Water Research and Education*, **155**, 43–52, https://doi.org/10.1111/j.1936-704X.2015.03194.x

African Development Bank (AFBD). United Nations Environmental Program (UNEP), GRID-Arendal, 2020 Sanitation and Wastewater Atlas of Africa. 284pp. https://www.afdb.org/sites/default/files/documents/publications/sanitation_and_wastewater_atlas_of_africa_compressed.pdf, accessed in June 2023.

Agrawal A. and Ribot J. (1999). Accountability in decentralization: a framework with South Asian and West African environmental cases. *The Journal of Developing Areas*, **33**, 473–502.

Akpan I. J., Soopramanien D. and Kwak D. H. (2020). Cutting-edge technologies for small business and innovation in the era of COVID-19 global health pandemic. Journal of Small Business and Entrepreneurship, 1–11, https://doi.org/10.1080/08276331.2020.1799294

Alhumoud J. M. and Madzikanda D. (2010) Public perceptions on water reuse options: the case of Sulaibiya wastewater treatment plant in Kuwait. *International Business and Economics Research Journal (IBER)*, **9**(1), https://doi.org/10.19030/iber.v9i1.515

Amoah P., Drechsel P. and Abaidoo R. C. (2005). Irrigated urban vegetable production in Ghana: sources of pathogen contamination and health risk elimination. Irrigation and Drainage, (54), S1–S118, https://doi.org/10.1002/ird.185

ANA (2024). Agências Infranacionais – Agência Nacional de Águas e Saneamento Básico (ANA) (www.gov.br)

Anderson J., Baggett S., Jeffrey P., McPherson L., Marks J. and Rosenblum R. (2008). Public acceptance of water reuse, Section 3 emerging topic, chapter 18. In: Water reuse: an international survey of current practice, issues and needs, B. Jiménez and T. Asano (eds.), IWA Publishing, UK, 628 pp. ISBN: 9781843390893, https://doi.org/10.2166/9781780401881

Andrade L., O'Dwyer J., O'Neill E. and Hynds P. (2018). Surface water flooding, groundwater contamination, and enteric disease in developed countries: a scoping review of connections and consequences. Environmental Pollution, 236, 540–549, https://doi.org/10.1016/j.envpol.2018.01.104

A/RES/70/1. United Nations (2015). Transforming our world: the 2030 Agenda for Sustainable Development. https://documents-dds-ny.un.org/doc/UNDOC/GEN/N15/291/93/PDF/N1529193.pdf, accessed September 2023.

Arsenault R., Bourassa C., Diver S., McGregor D. and Witham A. (2019). Including indigenous knowledge systems in environmental assessments: restructuring the process. Global Environmental Politics, 19(3), 120–132, https://doi.org/10.1162/glep_a_00519

Augsburg B. and Rodríguez-Lesmes P. (2020). Sanitation dynamics: toilet acquisition and its economic and social implications in rural and urban contexts. Journal of Water, Sanitation and Hygiene for Development, 10(4), 628–641, https://doi.org/10.2166/washdev.2020.098

Augsburg B. and Sainati T. (2020). WASH economics and financing: towards a better understanding of costs and benefits. Journal of Water, Sanitation and Hygiene for Development, 10(4), 615–617, https://doi.org/10.2166/washdev.2020.002

Austigard Å. D., Svendsen K. and Heldal K. K. (2018). Hydrogen sulphide exposure in wastewater treatment. Journal of Occupational Medicine and Toxicology, 13, 10, https://doi.org/10.1186/s12995-018-0191-z

Bagheri A., Fami H. S., Rezvanfar A., Asadi A. and Yazdani S. (2008). Perceptions of paddy farmers towards sustainable agricultural technologies: case of Haraz Catchments area in Mazandaran province of Iran. American Journal of Applied Sciences, 5(10), 1384–1391, https://doi.org/10.3844/ajassp.2008.1384.1391

Bargh J. A. and Barndollar K. (1996). Automaticity in action: the unconscious as repository of chronic goals and motives. In: The psychology of action: linking cognition and motivation to behavior, P. M. Gollwitzer and J. A. Bargh (eds.), The Guilford Press, pp. 457–481.

Bargh J. and Ferguson M. (2000). Beyond behaviorism: on the automaticity of higher mental processes. Psychological Bulletin, 126(6), 925–45, https://doi.org/10.1037/0033-2909.126.6.925. Corpus ID: 30847637.

Bayrau A., Boelee E., Drechsel P. and Dabbert S. (2010). Wastewater use in crop production in peri-urban areas of Addis Ababa: impacts on health in farm household. Environment and Development Economies, 16(1), 25–49.

Becerra-Castro C., Lopes A. R., Vaz-Moreira I., Silva E. F., Manaia C. M. and Nunes O. C. (2015). Wastewater reuse in irrigation: a microbiological perspective on implications in soil fertility and human and environmental health. Environment International, 75, 117–135, https://doi.org/10.1016/J.ENVINT.2014.11.001

Bernard K., Drechsel P., Seidu R., Amerasingue P., Cofie O. and Konradsen F. (2010). Harnessing farmers' knowledge and perceptions for health risk reduction in wastewater-irrigated agriculture. In: Wastewater irrigation and health, P. Drechsel, C. Scott, L. Raschid-Sally, M. Redwood and A. Bahri (eds.), Chapter 17, IDRC-IWMI, pp. 337–354.

Big Blue Ocean Clean up (2023). https://www.bigblueoceancleanup.org/news/2021/11/4/is-sewage-pollution-still-a-major-threat-to-our-oceans, accessed in November 2023.

BIO by Deloitte (2015). Cranfield University, Directorate-General for Environment (European Commission), ICF International, https://data.europa.eu/doi/10.2779/603205

Bisung E., Karanja D. M., Abudho B., Oguna Y., Mwaura N., Ego P. and … Elliott S. J. (2016). One community's journey to lobby for water in an environment of privatized water: is Usoma too poor for the pro-poor program? *African Geographical Review*, **35**(1), 70–82, https://doi.org/10.1080/19376812.2015.1088391

Borjas P. C. (2014). Public policy of sanitation: an analysis of recent Brazilian experience (Política pública de saneamento básico: uma análise da recente experiência brasileira). *Saúde e Sociedade, São Paulo*, **23**(2), 432–447, https://doi.org/10.1590/S0104-12902014000200007

Borja-Vega C., Grabinsky J. and Kløve E. (2022). Introducing a framework for analyzing weaknesses in institutional service delivery and the human rights to water and sanitation: case studies from the Democratic Republic of Congo, Haiti, Mozambique, and Niger. *Water*, **14**, 3209, https://doi.org/10.3390/w14203209

Branchet P., Arpin-Pont L., Piram A., Boissery P., Wong-Wah-Chung P. and Doumenq P. (2021). Pharmaceuticals in the marine environment: what are the present challenges in their monitoring? *Science of the Total Environment*, **766**, 142644, https://doi.org/10.1016/j.scitotenv.2020.142644

Brown C. and Heller L. (2017). Affordability in the provision of water and sanitation services: evolving strategies and imperatives to realise human rights. *International Journal of Water Governance*, **6**, 19–38. https://journals.open.tudelft.nl/ijwg/article/view/5812

Cabral B. G. C., Chernicharo C. A. L., Hoffmann H., Neves P. N. P., Platzer C., Bressani-Ribeiro T. and Rosenfeldt S. (2017). Resultados do Projeto de Medições de Biogás em Reatores Anaeróbios/Probiogás (Results of the biogas measurement project in anaerobic reactors/Probiogás); Deutsche Gesellschaf //efaidnbmnnnibpcajpcglclefindmkaj/; https://rotaria.net/wp-content/uploads/2023/10/PROBIOGAS-Resultados-ETEs.pdf, accessed in January 2024.

Cairns-Smith S., Hill H. and Nazarenke E. (2014). Urban sanitation: why a portfolio is needed. Boston Consulting Group, Boston, MA, USA. Available online: http://www.bcg.com/documents/file178928.pdf, accessed on 23 May 2016.

Campos C., Casas C., Bohórquez P., Cárdenas M., Cáceres L. and Muñoz M. (2001). Evaluation of water ecotoxicity tests in the municipality of Mosquera, Cundinamarca (Colombia). Quebec, Canadá. http://hdl.handle.net/10625/35803

Campos C., Valderrama L. and Venegas C. (2012). Talleres de Educación Ambiental. Pontificia Universidad Javeriana. Editorial Javeriana. Bogotá. **65** (in Spanish).

Caretta M. A. and Borjeson L. (2015). Local gender contract and adaptive capacity in smallholder irrigation farming: a case study from the Kenyan drylands. *Gender Place and Culture*, **22**(5), 644–661, https://doi.org/10.1080/0966369X.2014.885888

Caretta M. A., Mukherji A., Arfanuzzaman M., Betts R. A., Gelfan A., Hirabayashi Y., Lissner T. K., Liu J., Lopez Gunn E., Morgan R., Mwanga S. and Supratid S. (2022). Water. Review. In: B. E. Jimenez Cisneros and Z. Kundzewicz (eds.), for chapter Climate change 2022: impacts, adaptation and vulnerability. Contribution of working group II to the sixth assessment report of the intergovernmental panel on climate change, Portner H.-O., Roberts D. C., Tignor M., Poloczanska E. S., Mintenbeck K., Alegria A., Craig M., Langsdorf S., Loschke S., Moller V., Okem A., Rama B. (eds.), Cambridge University Press, Cambridge, UK and New York, NY, USA, pp. 551–712, https://doi.org/10.1017/9781009325844.006

Carr E. R. and Thompson M. C. (2014). Gender and climate change adaptation in agrarian settings: current thinking, new directions, and research frontiers. *Geography Compass*, 8(3), 182–197, https://doi.org/10.1111/gec3.12121

Carratala A., Bachmann V., Julian T. R. and Kohn T. (2020). Adaptation of human Enterovirus to warm environments leads to resistance against chlorine disinfection. *Environmental Science and Technology*, **54**(18), 11292–11300, https://doi.org/10.1021/acs.est.0c03199

Castleden H. E., Martin D., Cunsolo A., Harper S., Hart C., Sylvestre P., Stefanelli R., Day L. and Lauridsen K. (2017). Implementing indigenous and Western knowledge systems (part 2): 'You have to take a backseat' and abandon the arrogance of expertise. *The International Indigenous Policy Journal*, **8**, 4, https://doi.org/10.18584/iipj.2017.8.4.8

Castro de Esparza M. L., Sáenz R., Ratto A., Aurazo de Zumaeta M., Grados O., Ortiz D. *et al.* (1990). CEPIS/PAHO, IDRC. Assessment of health risks from the use of wastewater in agriculture.

Castro de Esparza M. L. and Aurazo de Zumaeta M. (2007). Informe final del proyecto AQUAtox: juventud, ciencia, salud y medio ambiente; y anexos. CEPIS/OPS-IDRC Canadá. Editor Organización Panamericana de la Salud. Dc. Colecciones, IDRC Research Results, Washington, DC, USA. http://hdl.handle.net/10625/36039 (2000–2009), https://idl-bnc-idrc.dspacedirect.org/items/b94319 87-db63-494e-919a-8e26a2911bfd

Cen T., Zhang X., Xie S. and Li D. (2020). Preservatives accelerate the horizontal transfer of plasmid-mediated antimicrobial resistance genes via differential mechanisms. *Environment International*, **138**(February), 105544, https://doi.org/10.1016/j.envint.2020.105544

Chen J. (2022). Green Fund: What it is, how it works, FAQs (investopedia.com), Green Fund, accessed in February 2024.

Chorus I. and Welker M. (eds.) (2021). Toxic cyanobacteria in water, 2nd edn, CRC Press, Boca Raton, FL, on behalf of the World Health Organization, Geneva, CH.

Cicogna M. P. V., Junior R. T., Gremaud A. P. and Figueiredo A. G. B. (2022). Financiamento do Saneamento: Linhas de Crédito e Perfil do Endividamento das Sociedades Anônimas no Brasil. *Revista Tempo do Mundo*, **29**. ago. 2022.

CONAGUA (2022). Situación del Subsector Agua Potable, Alcantarillado y Saneamiento, edición 2022, Secretaría de Medio Ambiente y Recursos Naturales, 150 pp., SGAPDS-13-22-a.pdf (www.gob.mx), accessed in August 2023 (in Spanish).

CONAGUA (2023). https://www.gob.mx/conagua, accessed in December 2023.

Coninck H., Revi A., Babiker M., Bertoldi P., Buckeridge M., Cartwright A., Dong W., Ford J., Fuss S., Hourcade J.-C., Ley D., Mechler R., Newman P., Revokatova A., Schultz S., Steg L. and Sugiyama T. (2018). Strengthening and implementing the global response. In: Global warming of 1.5°C. An IPCC special report on the impacts of global warming of 1.5°C above preindustrial levels and related global greenhouse gas emission pathways, in the context of strengthening the global response to the threat of climate change, sustainable development, and efforts to eradicate poverty, V. Masson-Delmotte, P. Zhai, H.-O. Portner, D. Roberts, J. Skea, P. R. Shukla, A. Pirani, W. Moufouma-Okia, C. Pean, R. Pidcock, S. Connors, J. B. R. Matthews, Y. Chen, X. Zhou, M. I. Gomis, E. Lonnoy, T. Maycock, M. Tignor and T. Waterfield (eds.), IPCC, Cambridge University Press, Cambridge.

Conway K., Lebu S., Heilferty K., Salzberg A. and Manga M. (2023). On-site sanitation system emptying practices and influential factors in Asian low- and middle-income countries: a systematic review. *Hygiene and Environmental Health Advances*, **6**, https://doi.org/10.1016/j.heha.2023.100050

Cookey P. E., Darnswasdi R. and Ratanachai C. (2016). Local people's perceptions of lake basin water governance performance in Thailand. *Ocean & Coast Management*, **120**, 11–28. https://doi.org/10.1016/j.ocecoaman.2015.11.015

Cookey P. E., Kugedera Z., Alamgir M. and Brdjanovic D. (2020). Perception management of non-sewered sanitation systems towards scheduled faecal sludge emptying behaviour change intervention. *Humanities and Social Sciences Communications*, **7**, 183, https://doi.org/10.1057/s41599-020-00662-0

Correia M. L. S. F., Esperdião F. and Melo R. L. (2020). Evolução das políticas públicas de saneamento básico do Brasil, do PLANASA ao PACSaneamento. 1216_1583448349_SEP_2020__Com_identificao__pdf_ide.pdf.

Costa N. R. (2023). Basic sanitation policy in Brazil: ideas, institutions and challenges in the twenty-first century. *Ciência and Saúde Coletiva*, **28**(9), 2595–2600. https://doi.org/10.1590/1413-81232023289.20432022.

Cotruvo J., Bridgers D., Cairns W., Jiménez Cisneros B., Cunliffe D., Davidson D., de Roda Husma A., Eaton A., Fawell J., Golmer K., LoPiccolo D. and Nam Ong C. (2013). Water recovery and reuse: guideline for safe application of water conservation methods in beverage production and food processing. *A Publication of the ILSI Research Foundation Center, Washington*, **71**.

Crocker J., Saywell D. and Bartram J. (2017). Sustainability of community-led total sanitation outcomes: evidence from Ethiopia and Ghana. *International Journal of Hygiene and Environmental Health*, **220**(3), 551–557, https://doi.org/10.1016/j.ijheh.2017.02.011

Crocker J., Fuente D. and Bartram J. (2021). Cost effectiveness of community led total sanitation in Ethiopia and Ghana. *International Journal of Hygiene and Environmental Health*, **232**, 113682, https://doi.org/10.1016/j.ijheh.2020.113682

Crook R. and Manor J. (1998). Democracy and decentralisation in South Asia and West Africa. Cambridge University Press, Cambridge, https://doi.org/10.1017/9780511607899

Cunha Marques R. (2010). Regulation of Water and Wastewater Services: An International Comparison, IWA Publishing, Vol. **9**, https://doi.org/10.2166/9781780401492

Cunha A. S. (2011). Saneamento Básico no Brasil: desenho institucional e desafios federativos. Textos para discussão (1565). IPEA (Instituto de Pesquisa Econômica Aplicada), Rio de Janeiro. ISSN 1415-4765, 25p, https://repositorio.ipea.gov.br/bitstream/11058/1338/1/TD_1565.pdf (in Portuguese).

Cutter S., Osman-Elasha B., Campbell J., Cheong S. M., McCormick S., Pulwarty R., Supratid S., Ziervogel G., Calvo E., Daud Mutabazi D., Arnall A., Arnold M., Linnerooth Bayer L., Bohle H. G., Emrich C., Hallegatte S., Koelle B., Oettle N., Polack E., Ranger N., Rist S., Suarez P. and Wilches-Chaux G. (2012). Managing the risks from climate extremes at the local level. In: Managing the risks of extreme events and disasters to advance climate change adaptation. A special report of working groups I and II of the intergovernmental panel on climate change, C. B. Field, V. Barros, T. F. Stocker, Q. Dahe, D. J. Dokken, K. L. Ebi, M. D. Mastrandrea, K. J. Mach, G. K. Plattner, S. K. Allen, M. Tignor and P. M. Midgley (eds.), Cambridge University Press, Cambridge, UK and New York, NY, USA, pp. 291–338.

Danert K. and Hutton G. (2020). Shining the spotlight on household investments for water, sanitation and hygiene (WASH): let us talk about HI and the three 'T's. *Journal of Water, Sanitation and Hygiene for Development*, **10**(1), 1–4, https://doi.org/10.2166/washdev.2020.139

Dash J. and Kapur D. (2021). Understanding effectiveness of capacity development: lessons from sanitation capacity building platform (SCBP), Part II: Sanitation capacity building platform: understanding the process and effectiveness, National Institute of Urban Affairs and Sanitation Capacity and Building Platform, chromeextension://efaidnbmnnnibpcajpcglclefindmkaj/, https://scbp.niua.org/sites/default/files/Capacity_Development.pdf

de França Doria M., Segurado P., Korc M., Heller L., Jimenez Cisneros B., Hunter P. R. and Forde M. (2021). Preliminary assessment of COVID-19 implications for the water and sanitation sector in Latin America and the Caribbean. *International Journal of Environmental Research and Public Health*, **18**, 11703, https://doi.org/10.3390/ijerph182111703

Delaire C., Peletz R., Haji S., Kones J., Samuel E., Easthope-Frazer A., Charreyron E., Wang E., Feng A., Mustafiz R., Jabeen Faria I., Antwi-Agyei P., Donkor E., Adjei K., Monney I., Kisiangani J., MacLeod C., Mwangi B. and Khush R. (2021). How much will safe sanitation for all cost? Evidence from five cities. *Environmental Science and Technology*, **55**(1), 767–777, https://doi.org/10.1021/acs.est.0c06348, pubs.acs.org/action/showCitFormats?doi=10.1021/acs.est.0c06348andref=pdf

Department of Environment and Natural Resources (2023). Republic of Philippines, https://www.denr.gov.ph/, accessed in December 2023.

De Shay R., Comeau D. L., Sclar G. D., Routray P. and Caruso B. A. (2020). Community perceptions of a multilevel sanitation behavior change intervention in rural Odisha, India. *International Journal of Environmental Research and Public Health*, **17**(12), 4472, https://doi.org/10.3390/ijerph17124472

Dickin S., Bayoumi M., Giné R., Andersson K. and Jiménez A. (2020). Sustainable sanitation and gaps in global climate policy and financing. *NPJ Clean Water*, **3**(1), 24, https://doi.org/10.1038/s41545-020-0072-8

Dickin S., Schuster-Wallace C., Qadir M. and Pizzacalla K. (2016). A review of health risks and pathways for exposure to wastewater use in agriculture. *Environmental health perspectives*, **124**(7), 900–909.

Dijksterhuis A. and van Knippenberg A. (1998). The relation between perception and behavior, or how to win a game of trivial pursuit. *Journal of Personality and Social Psychology*, **74**(4), 865–877, https://doi.org/10.1037/0022-3514.74.4.865

Djoudi H., Locatelli B., Vaast C., Asher K., Brockhaus M. and Basnett Sijapati B. (2016). Beyond dichotomies: gender and intersecting inequalities in climate change studies. *Ambio*, **45**(3), 248–262, https://doi.org/10.1007/s13280-016-0825-2

Dodane P. H., Mbeguere M., Sow O. and Strande L. (2012). Capital and operating costs of full-scale fecal sludge management and wastewater treatment systems in Dakar, Senegal. *Environmental Science and Technology*, **46**, 3705–3711, https://doi.org/10.1021/es2045234

Doménech L. (2011). Rethinking water management: from centralized to decentralized water supply and sanitation model. *Documents d'anàlisi geogràfica*, **57**(2), 593–310.

Dottori F., Szewczyk W., Ciscar Martinez J. C., Zhao F., Alfieri L., Hirabayashi Y., Bianchi A., Mongelli I., Frieler K., Betts R. and Feyen L. (2018). Increased human and economic losses from river flooding with anthropogenic warming. *Nature Climate Change*, **8**(9), 781–786, https://doi.org/10.1038/s41558-018-0257-z

Eakin H., Shelton R., Baeza A., Bojorquez-Tapia L., Flores S., Parajuli J., Grave I., Estrada A. and Hernandez B. (2020). Expressions of collective grievance as a feedback in multi-actor adaptation to water risks in Mexico City. *Regional Environmental Change*, **20**(1), 17, https://doi.org/10.1007/s10113-020-01588-8

Eawag (2021). Massive open online courses (MOOCs): MOOC series 'sanitation, water and solid waste for development'. Department of Sanitation, Water Solid Waste for Development. Swiss Federal Institute of Aquatic Science and Technology (Eawag). https://www.eawag.ch/en/department/sandec/e-learning/moocs/

EFSA (2024). European Food Safety Authority, https://www.efsa.europa.eu/en/topics/topic/vector-borne-diseases, accessed in February 2024.

Ehalt Macedo H., Lehner B., Nicell J., Grill G., Li J., Limtong A. and Shakya R. (2022). Distribution and characteristics of wastewater treatment plants within the global river network. *Earth System Science Data*, **14**, 559–577, https://doi.org/10.5194/essd-14-559-2022

Ellis A., Haver J., Villasenor J., Parawan A., Venkatesh M., Freeman M. C. and Caruso B. A. (2016). WASH challenges to girls' menstrual hygiene management in Metro Manila, Masbate, and South-Central Mindanao, Philippines. *Waterlines*, **35**(3), 306–323, https://doi.org/10.3362/1756-3488.2016.022

ESAWAS, Eastern and Southern Africa Water and Sanitation Regulators Association (2022). The water supply and sanitation regulatory landscape across Africa: continent-wide synthesis report, https://www.esawas.org/repository/Esawas_Report_2022.pdf, accessed in February 2024.

ESAWAS (2023). https://www.esawas.org/index.php/component/content/article/33-articles/102-defining-an-institutional-framework-for-cwis-doing-things-differently, accessed in December 2023.

Esquivel A. (2015). Acceptance and water reuse. An assessment of the water reuse program operating in San Marcos. Master of Public Administration. Texas State University, Texas, 86p.

European Commission (2010). Water reuse for irrigation. https://water.jrc.ec.europa.eu/wreuse.html, https://environment.ec.europa.eu/topics/water/water-reuse_en, accessed in November 2023.

European Commission, *et al.* (2015). Directorate-General for environment. In: Optimising water reuse in the EU – final report, part I, J. Knox, P. Jeffrey and L. Van Long (eds.), Publications Office. **2015**, Corporate author(s).

FAO (2017). Reuse of water for agriculture in Latin America and the Caribbean, state, principles and needs. Chile.

Fauconnier I., Jenniskens A., Perry P., Fanaian S., Sen S., Sinha V. and Witmer L. (2018). Women as change-makers in the governance of shared waters. UN Water, Gland, Switzerland. 50pp, https://doi.org/10.2305/IUCN.CH.2018.22.en

Fewtrell L. and Bartram J. (2001). Water quality. Guidelines, standards and health: assessment of risk and risk management for water-related infectious disease. IWAP and WHO, London. 431p. ISBN 1 900222 28 0 (IWA Publishing), ISBN 92 4 154533 X (World Health Organization).

Fielding K. S., Dolnicar S. and Schultz T. (2018). Public acceptance of recycled water. *International Journal of Water Resources Development*, **35**(4), 551–586, https://doi.org/10.1080/07900627.2017.1419125

Flores Uijtewaal B., Goksu A. and Saltiel B. (2018). Incentives for improving water supply and sanitation service delivery: a South American perspective. Water Global Practice: Knowledge World Bank, World Bank Document 126196-14-5-2018-12-2-53-WTex.

Freeman M. C., Garn J. V., Sclar G. D., Boisson S., Medlicott K., Alexander K. T., Penakalapati G., Anderson D. M., Mahtani A. G., Grimes and Classen T. (2017). The impact of sanitation on infectious disease and nutritional status: a systematic review and meta-analysis. *International Journal of Hygiene and Environmental Health*, **220**(6), 928–949, https://doi.org/10.1016/j.ijheh.2017.05.007

Garduño-Jiménez A. L. and Carter L. J. (2023). Insights into mode of action mediated responses following pharmaceutical uptake and accumulation in plants. *Frontiers in Agronomy*, **5**, 1293555. https://doi.org/10.3389/fagro.2023.1293555

Garduño-Jiménez A. L., Durán-Álvarez J. C., Ortori C. A., Abdelrazig S., Barrett D. A. and Gomes R. L. (2023). Delivering on sustainable development goals in wastewater reuse for agriculture: initial prioritization of emerging pollutants in the Tula Valley, Mexico. *Water Research*, **238**, 119903, https://doi.org/10.1016/j.watres.2023.119903

Gatto D., Salas Barboza A., Garcés V., Rodríguez Álvarez M. S., Iribarnegaray M. A., Liberal V. I., Fasciolo G. E., van Lier J. B. and Seghezzo L. (2015). The use of (treated) domestic wastewater for irrigation. Current situation and future challenges. *International Journal of Water and Wastewater Treatment*, **1**(2), 1–10, https://doi.org/10.16966/ijwwwt.107

Gaw S., Thomas K. V. and Hutchinson T. H. (2014). Sources, impacts and trends of pharmaceuticals in the marine and coastal environment. *Philosophical Transactions of the Royal Society B, Biological Science*, **369**, 20130572, https://doi.org/10.1098/rstb.2013.0572

GLASS (2021/2022). Assessment report 2021/2022, https://glaas.who.int/glaas/data, accessed in February 2024.

GLAAS (2022a). Progress on sanitation and hygiene in Africa 2000–2022. UN water global analysis and assessment of sanitation and drinking-water. UN Water, World Health Organization and UNICEF, https://www.unicef.org/media/147516/file/Africa-WASH-regional-snapshot-2022.pdf, accessed in February 2024.

GLAAS (2022b). ES WHO 2022a but UN-water also indicates it is their report. Strong systems and sound investments: evidence on and Key insights into accelerating progress on sanitation, drinking-water and hygiene. UN water global analysis and assessment of sanitation and drinking-water. UN Water and World Health Organization.

GMI (2013). Global Methane Initiative. Municipal Wastewater Methane: Reducing Emissions, Advancing Recovery and Use Opportunities. January 2013. Fact Sheet. 4p. https://www.globalmethane.org/documents/ww_fs_eng.pdf

Gomes A. R., Justino C., Rocha-Santos T., Freitas A. C., Duarte and Pereira R. (2017). Review of the ecotoxicological effects of emerging contaminants to soil biota. *Journal of Environmental Science and Health, Part A*, **52**(10), 992–1007, https://doi.org/10.1080/10934529.2017.1328946

Gonda N. (2016). Climate change, 'technology' and gender: 'adapting women' to climate change with cooking stoves and water reservoirs. *Gender, Technology and Development*, **20**(2), 149–168, https://doi.org/10.1177/0971852416639786

Governance Institute (2023). https://www.governanceinstitute.com.au/resources/what-is-governance/, accessed in December 2023.

Greenwood N. N. and Earnshaw A. (1997). Chemistry of the elements, 2nd ed., Butterworth-Heinemann. ISBN 978-0-08-037941-8.

Gu Q., Chen Y., Pody R., Cheng R., Zheng X. and Zhang Z. (2015). Public perception, and acceptability toward reclaimed water in Tianjin. *Resources, Conservation and Recycling*, **104**, 291–299, https://doi.org/10.1016/j.resconrec.2015.07.013

Gundy P. M., Gerba C. P. and Pepper I. L. (2009). Survival of coronaviruses in water and wastewater. *Food and Environmental Virology*, **1**, 10–, https://doi.org/10.1007/s12560-008-9001-6

Gupta J. and Pahl-Wostl C. (2013). Global water governance in the context of global and multilevel governance: its need, form, and challenges. *Ecology and Society*, **18**(4).

GWP (2000). Towards water security: a framework for action. Global Water Partnership, Stockholm, Sweden. Available at: www.gwpforum.or, accessed in November 2023.

GWP (2008). Sustainable sanitation and water management toolbox, GWP toolbox. Integrated Water Resources Management. https://sswm.info/node/1177, accessed in January 2024.

Hachimoto K. (2021). Institutional frameworks for onsite sanitation management systems. ADBI Development Case Study No. 2021-1 (June). Asian Development

Bank Institute. 18p. https://www.adb.org/sites/default/files/publication/711441/adbi-cs2021-01.pdf, accessed in January 2023.

Halpern J. and Trémolet S. (2006). Regulation of water and sanitation services: getting better service to poor people. GPOBA, World Bank, Washington, DC. https://documents.worldbank.org/en/publication/documents-reports/documentdetail/601751468166154983/regulation-of-water-and-sanitation-services-getting-better-service-to-poor-people, accessed in March 2025.

Harmsworth G., Shaun A. and Mahuru R. (2016). Indigenous Māori values and perspectives to inform freshwater management in Aotearoa-New Zealand. *Ecology and Society*, **21**(4), https://doi.org/10.5751/ES-08804-210409

Hartley T. (2006). Public perception and participation in water reuse. *Desalination*, **187**, 115–126, https://doi.org/10.1016/j.desal.2005.04.072

Heller L. (2020). Human rights and the privatization of water and sanitation services. Report of the Special Rapporteur on the human rights to safe drinking water and sanitation (A/75/208), https://www.ohchr.org/en/documents/thematic-reports/a75208-human-rights-and-privatization-water-and-sanitation-services, accessed in February 2024.

Heller L., Albuquerque C. D., Roaf V. and Jiménez A. (2020). Overview of 12 years of special rapporteurs on the human rights to water and sanitation: looking forward to future challenges. *Water (Switzerland)*, **12**(9), https://doi.org/10.3390/W12092598

Howard G. (2021). The future of water and sanitation: global challenges and the need for greater ambition. *AQUA–Water Infrastructure, Ecosystems and Society*, **70**(4), 438–448 and *Journal of Water Supply: Research and Technology-Aqua*. 70. 10.2166/aqua.2021.127

Hughes J., Cowper-Heays K., Olesson E., Bell R. and Stroombergen A. (2020). Impacts and implications of climate change on wastewater systems: a New Zealand perspective. *Climate Risk Management*, **31**, 100262, https://doi.org/10.1016/j.crm.2020.100262

Hulland K., Martin N., Dreibelbis R., DeBruicker Valliant J. and Winch P. (2015). What factors affect sustained adoption of safe water, hygiene, and sanitation technologies? A systematic review of literature. EPPI-Centre, Social Science Research Unit, UCL Institute of Education, University College London, London. ISBN: 978-1-907345-77-7, https://www.3ieimpact.org/sites/default/files/2019-01/srs_2-_factors_for_sustained_wash_adoption.pdf, accessed in February 2024.

Hutchings P., Johns M., Jornet D., Scott C. and Van den Bossche Z. (2018). A systematic assessment of the pro-poor reach of development bank investments in urban sanitation. *Journal of Water, Sanitation and Hygiene for Development*, **8**(3), 402–414, https://doi.org/10.2166/washdev.2018.147

Hutton G. (2013). Global costs and benefits of reaching universal coverage of sanitation and drinking-water supply. *Journal of Water and Health*, **11**(1), 1–12, https://doi.org/10.2166/wh.2012.105

Hutton G. and Haller L. (2004). Evaluation of the costs and benefits of water and sanitation improvements at the global level. World Health Organization. Water, Sanitation and Health Team. WHO/SDE/WSH/04.04, https://www.researchgate.net/publication/228594838_Evaluation_of_the_Costs_and_Benefits_of_Water_and_Sanitation_Improvements_at_the_Global_Level, accessed in November 2023.

Hutton G. and Varughese M. (2020). Global and regional costs of achieving universal access to sanitation to meet SDG target 6.2. UNICEF, New York, 26pp, https://www.unicef.org/media/90806/file/WashReports-CostsOfSanitation.pdf, accessed in June 2023.

Huynh P. T. A. and Resurreccion B. P. (2014). Women's differentiated vulnerability and adaptations to climate-related agricultural water scarcity in rural central Vietnam. *Climate and Development*, **6**(3), 226–237, https://doi.org/10.1080/17565529.2014.886989

IICA, Instituo Interamericano de Convenção para a Agricultura (2017) Semeando saberes, inspirando soluções: Boas Práticas na Convivência com o Semiárido/ Sento Sé, Cinthia. Brasília: IICA, 94 p. ISBN: 978-92-9248-659-4

IPCC (2018). Summary for policymakers. IPCC Special Report on the Impacts of Global Warming of 1.5°C, https://www.ipcc.ch/sr15/chapter/spm/, accessed in February 2024.

IUCN (2020). IUCN global standard for nature-based solutions: a user-friendly, framework for the verification, design and scaling up of NbS, 1st edn, IUCN, Gland, Switzerland, https://doi.org/10.2305/iucn.ch.2020.08.en

IWRM Action Hub (2023), https://iwrmactionhub.org/, accessed December 2023.

Jagai J. S., Li Q., Wang S., Messier K. P., Wade T. J. and Hilborn E. D. (2015). Extreme precipitation and emergency room visits for gastrointestinal illness in areas with and without combined sewer systems: an analysis of Massachusetts data, 2003–2007. *Environmental Health Perspective*, **123**(9), 873–879, https://doi.org/10.1289/ehp.1408971

James H. (2019). Women, water and 'wicked problems': community resilience and adaptation to climate change in Northern Pakkoku, Myanmar. In: Population, development, and the environment: challenges to achieving the sustainable development goals in the Asia Pacific, H. James (ed.), Springer Singapore, Singapore, pp. 215–225, https://doi.org/10.1007/978-981-13-2101-6_13

Jenkins M. W., Cumming O. and Cairncross S. (2015). Pit latrine emptying behavior and demand for sanitation services in Dar Es Salaam, Tanzania. *International Journal of Environmental Research Public Health* **12**, 2588–2611. https://doi.org/10.3390/ijerph120302588

Jiménez B. (2007). Helminths ova control in wastewater and sludge for agricultural reuse, in water reuse new paradigm towards integrated water resources management in encyclopaedia of biological, physiological and health sciences, water and health, Vol. II. In: Life support system, W. Grabow (ed.), EOLSS Publishers Co. Ltd.-UNESCO, Paris, France, pp. 429–449. ISBN: UNESCO 93-3-103999-7, https://www.susana.org/en/knowledge-hub/resources-and-publications/library/details/1497

Jiménez B. and Asano T. (2008). Water reclamation and reuse around the world. Section 1: world overview, chapter 1. In: Water reuse: an international survey of current practice issues and needs, B. Jiménez and T. Asano (eds.), IWA Publishing, London, UK, pp. 3–26. ISBN: 978-1843390893, https://doi.org/10.2166/9781780401881

Jiménez B. and Chávez A. (1998). Removal of helminth eggs in an advanced primary treatment with sludge blanket. *Environmental Technology*, **19**(11), 1061–1071. https://doi.org/10.1080/09593331908616764

Jiménez B., Barrios J., Mendez J. and Diaz J. (2004). Sustainable management of sludge in developing countries. *Water Science and Technology*, **49**(10), 251–258. https://doi.org/10.2166/wst.2004.0656

Jiménez B., Drechsel P., Kone D., Bahri A., Raschid-Sally L. and Qadir M. (2010). General wastewater, sludge and excreta use situation, Chapter 1. In: Wastewater irrigation and health: assessing and mitigating risks in low-income countries, Drechsel and Scott (eds.), Earthscan Press, London, UK, pp. 3–28. ISBN: 978-1-84407-795-3 (hardback), ISBN: 978-1-84407-796-0 (pbk), https://doi.org/10.4324/9781849774666

Jiménez A., Saikia P., Giné R., Avello P., Leten J., Liss Lymer B., Schneider K. and Ward R. (2020). Unpacking water governance: a framework for practitioners. *Water*, **12**(3), 827, https://doi.org/10.3390/w12030827

Jiménez Cisneros B. (1995). Wastewater reuse to increase soil productivity. *Water Science and Technology*, **32**(12), 173–180. https://doi.org/10.1016/0273-1223(96)00152-7 180

Jiménez-Cisneros B. (2001). La Contaminación Ambiental en México: Causas, Efectos y Tecnología Apropiada. Limusa. 926pp. ISBN: 968-18-6042-X. ISBN: 978-96-8-186042-4. México (Distribución en Latinoamérica) (in Spanish).

Jiménez-Cisneros B. (2011). Safe sanitation in low economic development areas. In: Treatise on water science, P. Wilderer (ed.), Academic Press, Oxford, vol. **4**, pp. 147–200, https://doi.org/10.1016/B978-0-444-53199-5.00082-8

Jiménez Cisneros B. E., Oki T., Arnell N. W., Benito G., Cogley J. G., Döll P., Jiang T. and Mwakalila S. S. (2014). Freshwater resources, Chapter 3. In: Climate change impacts, adaptation, and vulnerability. Part A, 1st edn. Global and sectoral aspects. Contribution of working group II to the fifth assessment report of the intergovernmental panel on climate change, C. B. Field, V. R. Barros, D. J. Dokken, K. J. Mach, M. D. Mastrandrea, T. E. Bilir, M. Chatterjee, K. L. Ebi, Y. O. Estrada, R. C. Genova, B. Girma, E. S. Kissel, A. N. Levy, S. MacCracken, P. R. Mastrandrea and L. L. White (eds.), Cambridge University Press, Cambridge, UK and New York, NY, USA, pp. 229–269. http://ipcc-wg2.gov/AR5/images/uploads/WGIIAR5-Chap3_FGDall.pdf

JMP (Joint Monitoring Programme) (2022). WHO-UNICEF. Progress on household drinking water, sanitation and hygiene 2000-2020: five years into the SDGs. Geneva: World Health Organization and United Nations Children's Fund; 2021, https://washdata.org/sites/default/files/2022-01/jmp-2021-washhouseholds-highlights.pdf, accessed 21 October 2022.

JMP (Joint Monitoring Programme) (2023). Progress on household drinking water, sanitation and hygiene 2000–2022: special focus on gender. United Nations Children's Fund (UNICEF) and World Health Organization (WHO), New York, https://cdn.who.int/media/docs/default-source/wash-documents/jmp-2023_layout_v3launch_5july_low-reswhowebsite.pdf?sfvrsn=c52136f5_3anddownload=true, accessed in February 2024.

Kar K. and Chambers R. (2008). Handbook on community-led total sanitation. Plan International UK; Institute of Development Studies at the University of Sussex, London, UK. bloodwater.org/content/uploads/2023/10/rc314.pdf

Kehoe P. and Chang T. (2018). Innovative water use in an urban setting. Waterlines, https://www.wateronline.com/doc/innovative-water-use-in-an-urban-setting-0001, accessed in December 2023.

Keraita B. and Drechsel P. (2012). Implementing non-conventional options for safe water reuse in agriculture in resource poor environments Ghana-agriculture. International Water Management Institute. EPA2012.

Keraita B., Drechsel P., Seidu R., Amerasinghe P., Cofie O. O. and Konradsen F. (2010). Harnessing farmers' knowledge and perceptions for health-risk reduction in wastewater-irrigated agriculture. In: Wastewater irrigation and health: assessing and mitigation risks in low-income countries, P. Drechsel, C. A. Scott, L. Raschid-Sally, M. Redwood and A. Bahri (eds.), Earthscan-IDRC-IWMI, UK, pp. 337–354. https://idrc-crdi.ca/en/book/wastewater-irrigation-and-health-assessing-and-mitigating-risk-low-income-countries#:~:text=, ISBN: 9781844077953.

Keraita B., Davila J. M. S., Drechsel P., Winkler M. and Medlicott K. (2015). Risk mitigation for wastewater irrigation systems in low-income countries: opportunities and limitations of the WHO guidelines. In: Alternative water supply systems, London, pp. 367–369. https://edoc.unibas.ch/36298/

Khan S. J., Deere D., Leusch F. D. L., Humpage A., Jenkins M. and Cunliffe D. (2015). Extreme weather events: should drinking water quality management systems adapt to changing risk profiles? *Water Research*, **85**, 124–136, https://doi.org/10.1016/j.watres.2015.08.018

Khan H. K., Rehman M. Y. A. and Malik R. N. (2020). Fate and toxicity of pharmaceuticals in water environment: an insight on their occurrence in south Asia. *Journal of Environmental Management*, **271**, 111030, https://doi.org/10.1016/j.jenvman.2020.111030

Kihila J., Mtei K. M. and Njau K. N. (2014). Wastewater treatment for reuse in urban agriculture; the case of Moshi municipality, Tanzania. *Physics and Chemistry of the Earth, Parts A/B/C*, **72–75**, 104–110, https://doi.org/10.1016/j.pce.2014.10.004

Kilobe B. M., Mdegela R. H. and Mtambo M. M. A. (2013). Acceptability of wastewater resource and its impact on crop production in Tanzania: the case of Dodoma, Morogoro and Mvomero wastewater stabilization ponds. *Kivukoni Journal*, **1**(2), 94–103, https://www.suaire.sua.ac.tz/server/api/core/bitstreams/d703d053-0c18-49a9-8731-db5656509644/content, accessed in February 2024.

Kingsley F. (2021). Sewer workers: health hazards of working in sewage and sanitation.

Kishimoto S., Steinfort L. and Petitjean O. (2020). *The future is public: Towards democratic ownership of services*. Transnational Institute (TNI). Amsterdam, The Netherlands: 258 p. https://www.tni.org/files/publication-downloads/futureispublic_online_def.pdf.

Kitano N. and Qu F. (2021). Five lessons for shaping policies and programs to accelerate urban sanitation in Asia: A case study on the Beijing Gaobeidian wastewater treatment plant. ADB (Asian Development Bank), Policy Brief No. 6 https://www.adb.org/sites/default/files/publication/734796/adbi-brief-five-lessons-shaping-policies-092421.pdf, consulted May 2024.

Knight L. D. and Presnell S. E. (2005). Death by sewer gas: case report of a double fatality and review of the literature. *American Journal of Forensic Medicine & Pathology*, **26**(2), 181–5. PMID: 15894856, https://doi.org/10.1097/01.paf.0000163834.87968.08

Kundzewicz Z. W., Mata L. J., Arnell N., Döll P., Jiménez B., Miller K., Oki T., Şen Z. and Shiklomanov I. (2008). The implications of projected climate change for freshwater resources and their management. *Hydrology Science Journal*, **53**(1), 3. https://doi.org/10.1623/hysj.53.1.3(Q1)

Larson A. M. Democratic decentralization in the forestry sector: lessons learned from Africa, Asia and Latin America. The Politics of Decentralization: Forests, Power and People, https://www.researchgate.net/publication/237787535_Democratic_Decentralisation_in_the_Forestry_Sector_Lessons_Learned_from_Africa_Asia_and_Latin_America

Lau J. T., Tsui H., Lau M. and Yang X. (2004). SARS transmission, risk factors, and prevention in Hong Kong. *Emerging Infectious Diseases*, **10**(4), 587–92, https://doi.org/10.3201/eid1004.030628. PMID: 15200846; PMCID: PMC3323085.

Lautze J., de Silva S., Giordano M. and Sanford L. (2011). Putting the cart before the horse: water governance and IWRM. *United Nations Sustainable Journal*, **35**(1), 1–8, https://doi.org/10.1111/j.1477-8947.2010.01339.x

Liberath Msaki G., Njaua K. N., Treydtec A. C. and Lyimoe T. (2022). Social knowledge, attitudes, and perceptions on wastewater treatment, technologies, and reuse in Tanzania. Water Reuse, (2), 223–241.

Lipscomb M. and Schechter L. (2018). Subsidies versus mental accounting nudges: harnessing mobile payment systems to improve sanitation. *Journal of Development Economics*, **135**, 235–254, https://ssrn.com/abstract=2991771r, https://doi.org/10.2139/ssrn.2991771, https://doi.org/10.1016/j.jdeveco.2018.07.007

Madikizela L. M., Tavengwa N. T. and Chimuka L. (2017). Status of pharmaceuticals in African water bodies: occurrence, removal and analytical methods. *Journal of Environmental Management*, **193**, 211–220, https://doi.org/10.1016/J. JENVMAN.2017.02.022

Marti E., Variatza E. and Balcazar J. L. (2014). The role of aquatic ecosystems as reservoirs of antibiotic resistance. *Trends in Microbiology*, **22**(1), 36–41, https://doi. org/10.1016/j.tim.2013.11.001

Massoud M. A., Tarhini A. and Nasr J. A. (2009). Decentralized approaches to wastewater treatment and management: applicability in developing countries. *Journal of Environmental Management*, **90**(1), 652–659, https://doi.org/10.1016/j. jenvman.2008.07.001

Maurer M., Rothenberger D. and Larsen T. A. (2005). Decentralised wastewater treatment technologies from a national perspective: at what cost are they competitive? *Water Supply*, **5**(6), 145–154, https://doi.org/10.2166/ws.2005.0059

Mayilla W., Keraita B., Ngowi H., Konradsen F. and Magayane F. (2017). Perceptions of using low-quality irrigation water in vegetable production in Morogoro, Tanzania. *Environment, Development and Sustainability*, **19**(1), 165–183, https://doi. org/10.1007/s10668-015-9730-2

MDS and SNS (2021). Ministério do Desenvolvimento Regional – MDR (Brasil). Secretaria Nacional de Saneamento – SNS. Panorama do Saneamento Básico no Brasil 2021/Secretaria Nacional de Saneamento do Ministério do Desenvolvimento Regional, Brasília/DF, **2021**. 223p. https://www.gov.br/cidades/pt-br/acesso-a-informacao/acoes-e-programas/saneamento/snis/produtos-do-snis/panorama-do-saneamento-basico-do-brasil, accessed in November 2023.

Mezzelani M., Gorbi S. and Regoli F. (2018). Pharmaceuticals in the aquatic environments: evidence of emerged threat and future challenges for marine organisms. *Marine Environmental Research*, **140**, 41–60, https://doi.org/10.1016/j. marenvres.2018.05.001

Michetti M., Raggi M., Guerra E. and Viaggi D. (2019). Interpreting farmers' perceptions of risks and benefits concerning wastewater reuse for irrigation: a case study in Emilia-Romagna (Italy). *Water*, **11**(1), 108, https://doi.org/10.3390/w11010108

Mills F., Blackett I. and Tayler K. (2014). Assessing on-site systems and sludge accumulation rates to understand pit emptying in Indonesia, 37th WEDC International Conference Hanoi, Vietnam, https://wedc-knowledge.lboro.ac.uk/resources/conference/37/Mills-1904.pdf, Sustainable Sanitation Services for All in a Fast Changing World, referred paper 1904, consulted May 2024.

Mills F., Willetts J., Evans B., Carrard N. and Kohlitz J. (2020). Costs, climate and contamination: three drivers for citywide sanitation investment decisions. *Frontiers in Environmental Science*, **8**, 130, https://doi.org/10.3389/fenvs.2020.00130

Missirian A. and Schlenker W. (2017). Asylum applications respond to temperature fluctuations. *Science*, **358**(6370), 1610–1614, https://doi.org/10.1126/science. aao0432, https://www.jstor.org/stable/26401164

Mo J., Guo J., Iwata H., Diamond J., Qu C., Xiong J. and Han J. (2022). What approaches should be used to prioritize pharmaceuticals and personal care products for research on environmental and human health exposure and effects? Environmental Toxicology Chemistry, https://doi.org/10.1002/etc.5520. Epub ahead of print. PMID: 36377688.

MoLGRD&C (Ministry of Local Government Rural Development and Cooperatives) (2017) Institutional and Regulatory Framework for Faecal Sludge Management (IRF-FSM). Government of the People's Republic of Bangladesh. https://ocw. un-ihe.org/pluginfile.php/4172/mod_resource/content/1/FSM%20Framework-Mega%20City%20Dhaka.pdf. Accessed 22 April 2020.

Murphy H. M., Corston-Pine E., Post Y. and McBean E. A. (2015). Insights and opportunities: challenges of Canadian first nations drinking water operators. *The International Indigenous Policy Journal*, **6**(3), https://doi.org/10.18584/iipj.2015.6.3.7

Narayan A. S. (2022). Planning Citywide Inclusive Sanitation. Doctoral Thesis, ETH Zurich, Zurich, Switzerland, published by EWAG, https://www.research-collection.ethz.ch/handle/20.500.11850/561822, accessed in January 2024.

Narayan A. S., Marks S. J., Meierhofer R., Strande L., Tilley E., Zurbrügg C. and Lüthi C. (2021). Advancements in and integration of water, sanitation, and solid waste for low- and middle-income countries. *Annual Review of Environment and Resources*, **46**, 193–219, https://doi.org/10.1146/annurev-environ-030620-042304

Nataraj S. K. (2022). Emerging Pollutants Treatment in Wastewater. CRC Press. 1st ed. https://doi.org/10.1201/9781003214786. 326 p.

Neves-Silva P., Braga J. G. and Heller L. (2023). Different positions in society, differing views of the world: the privatization of water and sanitation services in Minas Gerais, Brazil. *Frontiers in Sustainable Cities*, **5**, 1165872, https://doi.org/10.3389/frsc.2023.1165872

Ngoran S. D. and Xue X. (2015). Addressing urban sprawl in Douala, Cameroon: lessons from Xiamen integrated coastal management. *Journal of Urban Management*, **4**(1), 53–72, https://doi.org/10.1016/j.jum.2015.05.001

NOM-001-SEMARNAT (2006). Norma Oficial Mexicana (in Spanish) available at http://legismex.mty.itesm.mx/normas/ecol/semarnat001-2022_03.pdf, consulted May 2024.

NRC (2009). Contaminants, National Research Council (US) Committee on emergency and continuous exposure guidance levels for selected submarine, hydrogen sulfide. National Academies Press (USA), www.ncbi.nlm.nih.gov/books/NBK219913/

OECD (2009). Private sector participation in water infrastructures. OECD Check list for Publication. 132pp, https://doi.org/10.1787/9789264059221, https://www.oecd-ilibrary.org/finance-and-investment/private-sector-participation-in-water-infrastructure_9789264059221-en.March

OECD (2015). Water resources allocation: sharing risks and opportunities, OECD studies on water. OECD Publishing, Paris, https://doi.org/10.1787/9789264229631-en

OECD (2019). Making blended finance work for water and sanitation: unlocking commercial finance for SDG 6, https://www.oecd-ilibrary.org/sites/5efc8950-en/1/2/1/index.html?itemId=/content/publication/5efc8950-en&_csp_=6f524d6f7dc250ba913c88ad8727c82b&itemIGO=oecd&itemContentType=book, accessed in January 2024.

OECD (2020). States of fragility, https://peacerep.org/2020/11/01/psrp-research-informs-oecds-states-of-fragility-2020/, accessed in January 2024.

Our World in Data team (2023). 'Ensure access to water and sanitation for all', published online at OurWorldInData.org. Retrieved from https://ourworldindata.org/sdgs/clean-water-sanitation [online resource].

Özerol G., Vinke-de Kruijf J., Brisbois M. C., Casiano Flores C., Deekshit P., Girard C., Knieper C., Mirnezami S. J., Ortega-Reig M., Ranjan P., Schröder N. J. S. and Schröter B. (2018). Comparative studies of water governance: a systematic review. *Ecology and Society*, **23**(4), 43, https://doi.org/10.5751/ES-10548-230443

Palmer M. A., Liu J., Matthews J. H., Mumba M. and D'Odorico P. (2015). Manage water in a green way. *Science*, **349**, 584–585, https://doi.org/10.1126/science.aac7778

Parsons M., Nalau J. and Fisher K. (2017). Alternative perspectives on sustainability: indigenous knowledge and methodologies. *Challenges in Sustainability*, **5**(1), 7–14, https://doi.org/10.12924/cis2017.05010007

Passos F., Bressani-Ribeiro T., Rezende S. and Chernicharo C. A. L. (2020). Potential applications of biogas produced in small-scale UASB-based sewage treatment plants in Brazil. Energies, (13), 3356, https://doi.org/10.3390/en13133356

Pastor R. and Miglio R. (2013). Programa de Educación en Ciencia y tecnología del Agua para la Población Infanto-Juvenil del Perú (AQUAtech). Universidad Agraria de la Molina con la cooperación y financiamiento de AECI (España), http://www.lamolina.edu.pe/proyectos/proyecto_AQUAtech/componentes.htm

Pastor R., Cano A., Miglio R. and Castro de Esparza M. L. (2011). Increase awareness of the environmental risk of PPCPs in Latin American schools. 8th IWA International Conference on Water Reclamation & Reuse, 26–29 September 2011. Barcelona, España, http://www.lamolina.edu.pe/proyectos/proyecto_AQUAtech/divulgacion/poster/IWA

Peal A., Evans B., Blackett I., Hawkins P. and Heymans C. (2014a). Fecal sludge management: a comparative assessment of 12 cities. *Journal of Water, Sanitation and Hygiene for Development*, 4(4), 563–575. https://doi.org/10.2166/washdev.2014.026

Peal A., Evans B., Blackett I., Hawkins P. and Heymans C. (2014b). Fecal sludge management (FSM): analytical tools for assessing FSM in cities. *Journal of Water, Sanitation and Hygiene for Development*, 4(3), 371–383, https://doi.org/10.2166/washdev.2014.139

Pelling M., Kerr R., Biesbroek R., Caretta M., Cissé G., Costello M., Ebi K., Gunn E., Parmesan C. Schuster-Wallace C.J., Tirado M.C., van Alst M. and Woodward A. (2021). Synergies between COVID-19 and climate change impacts and responses. *Journal of Extreme Events*, 8(03), 2131002, https://doi.org/10.1142/S2345737621310023

Peña-Guzmán C., Ulloa-Sánchez S., Mora K., Helena-Bustos R., Lopez-Barrera E., Alvarez J. and Rodriguez-Pinzón M. (2019). Emerging pollutants in the urban water cycle in Latin America: a review of the current literature. *Journal of Environmental Management*, **237**, 408–423, https://doi.org/10.1016/j.jenvman.2019.02.100

Peters D. (2011). Building an institutional framework (WS), http://archive.sswm.info/print/1489?tid=491, accessed in July 2022.

Pokhrel N. and Adhikary S. (2017). Tapping the unreached: Nepal small towns water supply and sanitation sector projects – a sustainable model of service delivery. Asian Development Bank, https://www.adb.org/sites/default/files/institutional-document/363611/tapping-unreached-nepal.pdf, accessed in February 2021, https://doi.org/10.22617/TIM178754-2

Pontificia Universidad Javeriana y Universidad de La Salle (2021). Proyecto Soluciones Integrales para la Paz. Soluciones Centradas en Sistemas de Producción Integrada con Base Agroecológica Sostenible (SIPIBAS), 120pp (internal report in Spanish).

Post V. and Athreye V. (2016). Financing sanitation, the essence of public and private funding for sanitation. Financing Sanitation Paper Series #2. Waste and Finish Society, https://www.susana.org/_resources/documents/default/3-2439-7-1455032722.pdf, accessed in December 2023.

Prinz W. (1990). A common coding approach to perception and action. In: Relationships between perception and action, O. Neumann and W. Prinz (eds.), Springer-Verlag, Berlin, Germany, pp. 167–201.

Qadir M. (2022). Chapter 13. Potential of municipal wastewater for resource recovery and reuse. In: Water and climate change, T. M. Letcher (ed.), Elsevier, pp. 263–271. ISBN 9780323998758, https://doi.org/10.1016/B978-0-323-99875-8.00013-6

Qadir M., Drechsel P., Jiménez Cisneros B., Kim Y., Pramanik A. and Mehta P. (2020). Global and regional potential of wastewater as a water, nutrient and energy source. *Natural Resources Forum*, **44**(1), 40–51, https://doi.org/10.1111/1477-8947.12187

Radin M., Jeuland M., Wang H. and Whittington D. (2020). Benefit–cost analysis of community-led total sanitation: incorporating results from recent evaluations. *Journal of Benefit-Cost Analysis*, **11**(3), 380–417, https://doi.org/10.1017/bca.2020.6

Razak S. and Drechsel P. (2010). Cost-effectiveness analysis of interventions for diarrhoeal disease reduction among consumers of wastewater-irrigated lettuce in Ghana. International Water Management Institute: *RePEc:ags:iwmibc:127726*, https://doi.org/10.22004/ag.econ.127726

Resolution N. (154/2021). Regulatory agency for water and sewage services of Minas Gerais State, Brazil, ARSAE-MG, Resolution No. 154/2021, 28 June 2021 (Resolução ARSAE-MG N 154, de 28 de junho DE 2021 – ARSAE MG), www.arsae.mg.gov.br/2021/06/28/resolucao-arsae-mg-no-154-de-28-de-junho-de-2021/, accessed in October 2023.

Ribot J. (2002). Democratic decentralization of natural resources: institutionalizing popular participation. World Resources Institute, Washington, DC, https://www.fao.org/3/XII/0775-A4.htm, accessed in November 2024.

Richmond M., Upadhyaya N. and Pastor A. O. (2021). An Analysis of Urban Climate Adaptation Finance. A Report from the Cities Climate Finance Leadership Alliance, https://www.climatepolicyinitiative.org/wp-content/uploads/2021/02/An-Analysis-of-Urban-Climate-Adaptation-Finance.pdf, accessed in December 2023.

Rigaud K., de Sherbinin K., Jones A., Bergmann B., Clement J., Ober V., Schewe K., Adamo J., McCusker S., Heuser B. and Midgley S. (2018). Groundswell: preparing for internal climate migration. World Bank, Washington, DC, http://hdl.handle.net/10986/29461, License: CC BY 3.0 IGO.

Ritchie H., Spooner F. and Roser M. (2019). 'Sanitation'. Published online at OurWorldInData.org. Retrieved from: https://ourworldindata.org/sanitation [online resource], accessed in November 2023.

Rivera F., Warren A., Ramirez E., Decamp O., Bonilla P., Gallegos E., Calderón A. and Sánchez J. T. (1995). Removal of pathogens from wastewaters by the root zone method (RZM) author links open overlay. *Water Science and Technology*, **32**(3), 211–218, https://doi.org/10.2166/wst.1995.0143

Robinson K., Robinson C. and Hawkins S. (2005). Assessment of public perception regarding wastewater reuse. *Water Supply*, **5**(1), 59–65, https://doi.org/10.2166/ws.2005.0008

Rodríguez D. J., Serrano H. A., Delgado A., Nolasco D. and Saltiel G. (2020). From waste to resource: shifting paradigms for smarter wastewater interventions in Latin America and the Caribbean. World Bank Group, USA, https://policycommons.net/artifacts/1262506/from-waste-to-resource/1835944/, accessed in February 2024. CID: 20.500.12592/5n0x06.

Rodríguez-Rodríguez G., Ballesteros H. M., Martínez-Cabrera H., Vilela R., Pennino M. G. and Bellido J. M. (2021). On the role of perception: understanding stakeholders' collaboration in natural resources management through the evolutionary theory of innovation. *Sustainability*, **13**, 3564, https://doi.org/10.3390/su13063564

Rogers P. and Hall A. (2003). Effective water governance. Tec background paper No. 7. Global Water Partnership Technical Committee, ISBN: 91-974012-9-3, https://www.gwp.org/globalassets/global/toolbox/publications/background-papers/07-effective-water-governance-2003-english.pdf

Saad D., Byrne D. and Drechsel P. (2017). Social perspectives on the effective management of wastewater. In: Physico-chemical wastewater treatment and resource recovery,

R. Farooq and Z. Ahmad (eds.), IntechOpen, Chapter 12, https://doi.org/10.5772/67312

Sainati T., Zakaria F., Locatelli G., Sleigh P. A. and Evans B. (2020). Understanding the costs of urban sanitation: towards a standard costing model. *Journal of Water, Sanitation and Hygiene for Development*, **10**(4), 642–658, https://doi.org/10.2166/washdev.2020.093

Saleem M., Burdet T. and Heaslip V. (2019). Health and social impacts of open defecation on women. *BMC Public Health*, **19**(1), 158. https://pubmed.ncbi.nlm.nih.gov/30727975, https://doi.org/10.1186/s12889-019-6423-z

Sanitation and water for all (2021). Reach out and reach up: insights into global perspectives on water, sanitation and hygiene, https://www.sanitationandwaterforall.org/reach-out-and-reach-global-study-external-perceptions-water-sanitation-and-hygiene, accessed in December 2023.

Santos F. F. S., Filho J. D., Machado C. T., Vasconcelos J. F. and Feitosa F. R. S. (2018). O desenvolvimento do saneamento básico no Brasil e as consequências para a saúde pública. *Revista Brasileira de Meio Ambiente*, **4**(1). Corrente, Piauí, Instituto Federal de Educação, Ciência e Tecnologia, https://revistabrasileirademeioambiente.com/index.php/RVBMA/article/view/127

Sayer J. A., Elliott C., Barrow E., Gretzinger S., Maginnis S., McShane T. and Shepherd G. (2004). The implications for biodiversity conservation of decentralized forest resources management, paper prepared on behalf of IUCN and WWF for the UNFF inter-sessional workshop on decentralisation Interlaken. Switzerland, May 2004, chrome-extension://efaidnbmnnnibpcajpcglclefindmkaj/, https://www.cifor.org/publications/pdf_files/events/documentations/interlaken/papers/JSayer.pdf, accessed in December 2023.

Schertenleib R., Lüthi C. C., Panesar A., Büürma M., Kapur D., Narayan A. S., Pres A., Salian P., Spuhler D. and Tempel A. (2021). A sanitation journey – principles, approaches and tools for urban sanitation. SuSanA@ Deutsche Gesellschaft für Internationale Zusammenarbeit (GIZ) GmbH and Eawag, Bonn, Germany and Dübendorf, Switzerland, 80pp, www.susana.org/en/knowledge-hub/resources-and-publications/library/details/4087

Schuster-Wallace C. J. and Dickson S. (2017). Pathways to a water secure community. In: Individuals and communities: the human face of water in the water security in a New World, Z. Adeel, R. Sandford and D. Devlaeminck (eds.), Springer, pp. 197–216, https://doi.org/10.1007/978-3-319-50161-1_10

Schuster-Wallace C. J., Dickin S. and Metcalfe C. (2014). Waterborne and foodborne diseases, climate change impacts on health. In Global Environmental Change, pp. 615–622, https://doi.org/10.1007/978-94-007-5784-4_102

Schuster-Wallace C. J., Metcalfe C. and Cave K. (2017). From waste to wealth: sustainable wastewater management in Uganda: wastewater management framework. United Nations University Institute for Water, Environment, and Health, https://inweh.unu.edu/wp-content/uploads/2021/03/Wastewater-Management-Framework_FINAL.pdf, accessed in December 2023.

Sclar G., Garn J., Penakalapati G., Alexander K., Krauss J., Freeman M., Boisson S., Medlicott K. and Clasen T. (2017). Effects of sanitation on cognitive development and school absence: a systematic review. *International Journal of Hygiene and Environmental Health*, **220**, 917–927, https://doi.org/10.1016/j.ijheh.2017.06.010

Sclar G., Penakalapati B. A., Caruso E. A., Rehfuess J. V., Garn K. T., Alexander M. C., Freeman S., Boisson K., Medlicott and Clasen T. (2018). Exploring the relationship between sanitation and mental and social well-being: a systematic review and qualitative synthesis. *Social Science & Medicine*, **217**, 121–134, https://doi.org/10.1016/j.socscimed.2018.09.016

Sclar G. D., Routray P., Majorin F., *et al.* (2022). Mixed methods process evaluation of a sanitation behavior change intervention in rural Odisha, India. *Global Implementation Research and Applications*, **2**, 67–84, https://doi.org/10.1007/s43477-022-00035-6

Scott D. and Becken S. (2010). Adapting to climate change and climate policy: progress, problems, and potentials. *Journal of Sustainable Tourism*, **18**(3), 283–295, https://doi.org/10.1080/09669581003668540

SEP (2024). https://www.google.com/search?q=la+escuela+es+nuestra+resultados&oq=la+escuela+es+nuestra+resultados&gs_lcrp=EgZjaHJvbWUyBggAEEUYOTIICAEQABgWGB7SAQg1MjczajBqNKgCALACAA&sourceid=chrome&ie=UTF-8, accessed in December 2023.

Seppala O. (2002). Effects of water and sanitation policy reform implementation: need for systemic approach and stakeholder participation. *Water Policy*, **4**(4), 367–388, https://doi.org/10.1016/S1366-7017(02)00036-3

SFPUC (2023). Onsite water reuse program guidebook. A guide for implementing onsite water reuse systems in San Francisco, //efaidnbmnnnibpcajpcglclefindmkaj/, https://sfpuc.org/sites/default/files/documents/OnsiteWaterReuseGuideAugust2022.pdf, accessed in December 2023.

Sinharoy S. S. and Caruso B. A. (2019). On world water day, gender equality and empowerment require attention. *The Lancet Planetary Health*, **3**(5), e202–e203, https://doi.org/10.1016/S2542-5196(19)30021-X

Somerville M. (2014). Developing relational understandings of water through collaboration with indigenous knowledges. *WIREs Water*, **1**(4), 401–411, https://doi.org/10.1002/wat2.1030

Speich B., Croll D., Fürst T., Utzinger J. and Keiser J. (2016). Effect of sanitation and water treatment on intestinal protozoa infection: a systematic review and meta-analysis. *The Lancet Infectious Diseases*, **16**, 87–99, https://doi.org/10.1016/S1473-3099(15)00349-7

Stephan M., Marshall G. and McGinnis M. (2019). An introduction to polycentricity and governance. In: Governing complexity, A. Thiel, W. Blomquist and D. Garrick (eds.), Cambridge University Press, New York, Chapter 1, pp. 21–44, https://doi.org/10.1017/9781108325721.002

Strande L., Ronteltap M. and Brdjanovic D. (2014). Faecal sludge management – systems approach for implementation and operation. IWA Publishing, London, UK, https://doi.org/10.2166/9781780404738, ISBN electronic: 781780404738.

Strande L., Schoebitz L., Bischoff F., Ddiba D., Okello F., Englund M., Warda B. J. and Niwagaba C. B. (2018). Methods to reliably estimate faecal sludge quantities and qualities for the design of treatment technologies and management solutions. *Journal of Environmental Management*, **223**, 898–907, https://doi.org/10.1016/j.jenvman.2018.06.100

Sultana F. (2018). Gender and water in a changing climate: challenges and opportunities. In: Water security across the gender divide, C. Frohlich, G. Gioli, R. Cremades and H. Myrttinen (eds.), Springer International Publishing, Cham, pp. 17–33, https://doi.org/10.1007/978-3-319-64046-4_2

Sunshine Coast Regional District (2019). Wastewater service review and asset management plans. Sunshine Coast Regional District, https://www.scrd.ca/wp-content/uploads/2023/01/2019-Wastewater-Service-Review-and-Asset-Management-Plan.pdf, accessed in December 2023.

SuSanA (2008). Towards more sustainable sanitation solutions – SuSanA Vision Document. Sustainable Sanitation Alliance (SuSanA), https://www.susana.org/en/knowledge-hub/resources-and-publications/library/details/267, accessed in November 2023.

Świacka K., Maculewicz J., Kowalska D., Caban M., Smolarz K. and Świeżak J. (2022). Presence of pharmaceuticals and their metabolites in wild-living aquatic organisms – current state of knowledge. *Journal of Hazardous Materials*, **424**, 127350, https://doi.org/10.1016/j.jhazmat.2021.127350

SWIM-SM (2013). Sustainable Water Integrated Management – Support Mechanism Contract Reference ENPI/2010/255-560 Workplan for the 2nd Year of Project Implementation September 2012–August 2013, Presented at the SWIM-SM Steering Committee Meeting 17–18 October 2012, Brussels, https://www.swim-sm.eu/files/SWIM-SM_2nd_PSC_WorkPlan_for_2013.pdf

Swiss Re (2024). https://www.swissre.com/our-business/public-sector-solutions/insurance-to-protect-and-enable-nature-based-solutions.html, accessed in February 2024.

Tang J. W., Marr L. C., Li Y. and Dancer S. J. (2021). COVID-19 has redefined airborne transmission. *BMJ*, **373**, n913, https://doi.org/10.1136/bmj.n913

Thebo A. L., Drechsel P., Lambin E. F. and Nelson K. L. (2017). A global, spatially-explicit assessment of irrigated croplands influenced by urban wastewater flows. *Environmental Research Letters*, **12**, 074008, https://doi.org/10.1088/1748-9326/aa75d1

Tiwari R. (2008). Occupational health hazards in sewage and sanitary workers. *Indian Journal of Occupational and Environmental Medicine*, **12**, 112–5, https://doi.org/10.4103/0019-5278.44691

Toilet Board Coalition (2019). Scaling up the sanitation economy 2020–2025. Toilet Board Coalition, https://www.toiletboard.org/media/52-Scaling_the_Sanitation_Economy.pdf (accessed in December 2023).

Toilet Board Coalition (2020). Sanitation economy markets: Kenya. Toilet Board Coalition, https://www.toiletboard.org/wp-content/uploads/2021/03/2020-Sanitation-Economy-Markets-Kenya-2020.pdf (accessed in December 2023).

Tsillas V. (2015). Research on water disputes. *Innovative Energy Research*, **4**(3) (omicsonline.org) https://www.omicsonline.org/open-access/research-on-water-disputes-ier-1000124.php?aid=69683, https://doi.org/10.4172/2576-1463.1000124

UN (2014). Every dollar invested in water, sanitation brings four-fold return in costs – UN, https://news.un.org/en/story/2014/11/484032, accessed in February 2024.

UNDESA (2015). https://sdgs.un.org/es/goals/goal6, accessed in January 2024.

UNDSSA – United Nations Department of Social and Social Affairs (2015). Transforming our world: the 2030 agenda for sustainable development. UNDSSA, Geneva, https://sustainabledevelopment.un.org/content/documents/21252030%20Agenda%20for%20Sustainable%20Development%20web.pdf

UN ESCAP – United Nations Economic and Social Commission for Asia and the Pacific (ESCAP), United Nations Human Settlements Program (UN-Habitat) and Asian Institute of Technology (AIT) (2015). Policy guidance manual on wastewater management with a special emphasis on decentralized wastewater treatment systems. Bangkok, Thailand. **144**. ISBN: 978-974-8257-87-7, https://repository.unescap.org/bitstream/handle/20.500.12870/75/ESCAP-2015-MN-Policy-guidance-manual-on-wastewater-management.pdf?sequence=1&isAllowed=y, accessed in February 2024.

UNGA (2010) resolution A/RES/64/292 available at: https://documents.un.org/doc/undoc/gen/n09/479/35/pdf/n0947935.pdf?token=UdxYkQ7EIg9f380c74&fe=true, consulted June 2024.

UN-Habitat and WHO (2021). Progress on wastewater treatment – global status and acceleration needs for SDG indicator 6.3.1. United Nations Human Settlements Programme (UN-Habitat) and World Health Organization (WHO), Geneva.

UNHRC – United Nations Human Rights Council (2010). Resolution on the human right to safe drinking water and sanitation. UNHRC, Geneva. Resolution A/HRC/RES/15/9.

UNICEF (2019). Guidance on menstrual health and hygiene. United Nations Children's Fund, New York, https://www.unicef.org/cmedia/91341/file/UNICEF-Guidance-menstrual-health-hygiene-2019.pdf, accessed in 21 October 2022.

UNICEF and WHO (2020). State of the world's sanitation: an urgent call to transform sanitation for better health, environments, economies and societies. United Nations Children's Fund (UNICEF) and the World Health Organization, New York, https://www.who.int/publications/i/item/9789240014473

UN-Water (2015a). Water and Sustainable Development. From Vision to Action. Report of the 2015 UN-Water Zaragoza Conference. 68, https://www.unwater.org/sites/default/files/app/uploads/2017/05/WaterandSD_Vision_to_Action-2.pdf, accessed in February 2024.

UN-Water (2015b). African Development Bank, UN, GRID-Arendal, 2020.

UN-Water (2024). Financing water and sanitation, Facts and Figures, https://www.unwater.org/water-facts/financing-water-and-sanitation, accessed in February 2024.

USDA (2024). https://www.usda.gov/oce/energy-and-environment/markets/water-quality, consulted May 2024.

US-EPA (United States Environmental Protection Agency) (2012a). Guidelines for water reuse. EPA/600/R-618. Washington, DC, 643, https://www.epa.gov/sites/default/files/2019-08/documents/2012-guidelines-water-reuse.pdf

US-EPA (2012b). Global Anthropogenic Emissions of Non-CO_2 Greenhouse Gases: 1990–2030 (EPA 430-R-12-006), http://www.epa.gov/climatechange/EPAactivities/economics/nonco2projections.html

Ventola C. L. (2015). Antibiotic resistance crisis, part 1: causes and threats. *International Journal of Medicine in Developing Countries*, **40**(4), 561–564, https://doi.org/10.24911/ijmdc.51-1549060699

Wang X., Li J., Guo T., Zhen B., Kong Q., Y., B., Li Z., Song N., Jin M., Xiao W., Zhu X. M., Gu C. Q., Yin J., Wei W., Yao W., Liu C., Li J. F., Ou G. R., Wang M. N., Fang T. Y., Wang G. J., Qiu Y. H., Wu H. H., Chao F. H. and Li J. W. (2005) Concentration and detection of SARS coronavirus in sewage from Xiao Tang Shan Hospital and the 309th Hospital of the Chinese People's Liberation Army. *Water Science and Technology*, **52**, 213–221, https://doi.org/10.2166/wst.2005.0266

Watt M., Watt S. and Seaton A. (1997). Episode of toxic gas exposure in sewer workers. *Occupational and Environmental Medicine*, **54**, 277–280. oenvmed00088-0061.pdf (nih.gov), https://doi.org/10.1136/oem.54.4.277. PMID: 9166135; PMCID: PMC1128703.

Werneck B., Vieira G. C. and Youssef L. M. (2020). Brazilian new basic sanitation law: the reform and its implication for new investments. Mayer Brown. 3p. https://www.mayerbrown.com/-/media/files/perspectives-events/publications/2020/09/the-legal-reform-of-brazilian-basic-sanitation-act.pdf?rev=66e855d178824d43ae6347bddc078623, accessed in February 2024.

WHO – World Health Organization (2006). Guidelines for the safe use of wastewater, excreta and greywater, Volume 1. World Health Organization. Geneve.114.

WHO – World Health Organization (2018). Guidelines on sanitation and health. World Health Organization, Geneva, 2018. Licence: CC BY-NC-SA 3.0 IGO.

WHO – World Health Organization (2020). Water, sanitation, hygiene and waste management for COVID-19. Interim Guidance. Available online: https://www. WHO/2019-nCoV/IPC_WASH/2020.3, accessed 2 May 2020.

WHO – World Health Organization (2020). United Nations Development Program. United Nations Environment. United Nations International Children Emergency Fund. State of World Sanitation. UNICEF and WHO, New York. 94.

WHO (2022). A strong systems and sound investments evidence on and key insights into accelerating progress on sanitation, drinking-water and hygiene, UN-Water Global Analysis and Assessment of Sanitation and Drinking-Water GLAAS 2022 REPORT World Health Organization; 2022. Licence: CC BY-NC-SA 3.0 IGO. https://www. who.int/publications/i/item/9789240065031, accessed in February 2024.

WHO and UNICEF (2021). Progress on household drinking water, sanitation and hygiene 2000–2020. Five years into the SDGs. WHO and UNICEF, New York, https://www. who.int/images/default-source/departments/wash/jmp21/en_who_wash_social-media_12102021_2.jpg?sfvrsn=be49be8a_36, accessed in February 2024.

WHO/CED/PHE/WSH (2018). Water sanitation and hygiene strategy, 2018–2025.

WHO-HEP,ECH,EHD (2022).World Health Organization;WHO/HEP/ECH/EHD/22.01. Compendium of WHO and other UN guidance on health and environment, 2022 update, https://www.who.int/publications/i/item/WHO-HEP-ECH-EHD-22.01

WHO-UNICEF JMP (2016). Sanitation, Sanitation | JMP (washdata.org), accessed in February 2024.

WHO-UNICEF JMP (2019). Estimates on the use of water, sanitation and hygiene by country, 2000–2017. April 2019. Available online: www.washdata.org, accessed on 10 May 2020.

Wikipedia (2024). https://en.wikipedia.org/wiki/ Sanitation, accessed in January 2024.

Wilson N. J. (2019). 'Seeing water like a state?': indigenous water governance through Yukon first nation self-government agreements. *Geoforum*, **104**, 101–113, https://doi.org/10.1016/j.geoforum.2019.05.003

Wolf J., Prüss-Ustün A., Cumming O., Bartram J., Bonjour S., Cairncross S., Clasen T., Colford J. M., Curtis V. and De France J. (2014). Systematic Review: Assessing the impact of drinking water and sanitation on diarrhoeal disease in low- and middle-income settings: systematic review and meta-regression. *Tropical Medicine. International Health*, **19**, 928–942, https://doi.org/10.1111/tmi.12331

World Bank (2003). World Development Report 2004: Making Services Work for Poor People, https://hdl.handle.net/10986/5986. License: CC BY 3.0 IGO.

World Bank (2017). Easing the transition to commercial finance for sustainable water and sanitation. World Bank, Washington, DC, https://www.oecd.org/water/Background-Document-OECD-GIZ-Conference-Easing-the-Transition.pdf, accessed in February 2024.

World Bank (2019a). Health, safety and dignity of sanitation workers: an initial assessment. World Bank Group, Washington, DC, https://www.who.int/publications/m/item/health-safety-and-dignity-of-sanitation-workers, accessed in December 2023.

World Bank (2019b). Women in water utilities: breaking barriers. World Bank Group, Washington, DC, chrome extension://efaidnbmnnnibpcajpcglclefindmkaj/, https://www.ib-net.org/docs/Women%20in%20Water%20Utilities%20Breaking%20Barriers.pdf, accessed in February 2024.

World Bank (2023a). Sanitation overview: development news, research, data. Available from: https://www.worldbank.org/en/topic/sanitation, accessed in December 2023.

World Bank, WB (2023b). 5 Trends in public–private partnerships in water supply and sanitation, https://ppp.worldbank.org/public-private-partnership/5-trends-public-private-partnerships-water-supply-and-sanitation

World Health Organization, WHO (2006). WHO guidelines for the safe use of wastewater, excreta and greywater, Vol. **1**, World Health Organization, France, 114. https://www.who.int/publications/i/item/9241546824, accessed in February 2024.

World Health Organization (2018). Guidelines for sanitation and health. Organización Mundial de la Salud, Ginebra, **2019**. Licence: CC BY-NC-SA 3.0 IGO. ISBN: 978 92 4 151470 5, https://www.who.int/publications/i/item/9789241514705

World Health Organization (WHO) (2021). COVID-19 weekly epidemiological update. 7 March 2021. Available online: https://www.who.int/publications/m/item/weekly-operational-update-on-covid-19 (accessed on 7th June 2021).

World Health Organization (WHO). COVID-19 weekly epidemiological update. 7 March 2021. Available online: https://www.who.int/publications/m/item/weekly-operational-update-on-covid-19, accessed in June 2022.

World Health Organization and the United Nations Children's Fund (UNICEF) (2020). Water, sanitation, hygiene, and waste management for SARS-CoV-2, The virus that causes COVID-19: Interim Guidance. Available online: https://www.who.int/publications/i/item/WHO-2019-nCoV-IPC-WASH-2020.4, September 2023 (accessed on 3rd August 2020).

World Health Organization. United Nations Development Program. United Nations Environment. United Nations International Children Emergency Fund (2020). State of world sanitation. UNICEF and WHO, New York, **94**, https://www.unicef.org/reports/state-worlds-sanitation-2020, accessed in February 2024.

WSP (2011). Economic impacts of inadequate sanitation in India. Water and Sanitation Programme, The World Bank, http://documents.worldbank.org/curated/en/820131468041640929/Economic-impacts-of-inadequate-sanitation-in-India, accessed in February 2024.

Wu X., Schuyler House R. and Peri R. (2016). Public–private partnerships (PPPs) in water and sanitation in India: lessons from China. *Water Policy*, **18**(S1), 153, https://doi.org/10.2166/wp.2016.010

WWAP, United Nations World Water Assessment Programme (2015). The United Nations world water development report 2015: water for a sustainable world. UNESCO, Paris, https://www.unwater.org/publications/un-world-water-development-report-2015, accessed in February 2024.

WWAP, United Nations World Water Assessment Programme (2017). The United Nations world water development report 2017. Wastewater: The Untapped Resource, Paris. ISBN 978-92-3-100201-4, https://www.unwater.org/publications/un-world-water-development-report-2017

Xue W., Li X., Yang Z. and Wei J. (2022). Are house prices affected by $PM_{2.5}$ pollution? Evidence from Beijing, China. *International Journal of Environmental Research and Public Health*, **19**(14), 8461, https://doi.org/10.3390/ijerph19148461. PMID: 35886314; PMCID: PMC9317985.

Yadav S. S. and Lal R. (2018). Vulnerability of women to climate change in arid and semi-arid regions: the case of India and South Asia. *Journal of Arid Environments*, **149**, 4–17, https://doi.org/10.1016/j.jaridenv.2017.08.001

Yishay A. B., Fraker A., Guiteras R., Palloni G., Shah N. B., Shirrell S. and Wan P. (2017). Microcredit and willingness to pay for environmental quality: evidence from a randomized-controlled trial of finance for sanitation in rural Cambodia. *Journal of Environmental Economics and Management*, **86**, 121–140, https://doi.org/10.1016/j.jeem.2016.11.004

Yuliani E. (2004). Decentralization, deconcentration and devolution: what do they mean? https://www.cifor.org/publications/pdf_files/interlaken/Compilation.pdf, accessed in October 2023.

IWA Publishing's authorised EU representative for General Product
Safety Regulations is Diane D'Arras, 15 rue Duret, 75116 Paris,
France, e-mail: safety@iwap.co.uk.

Printed and bound by CPI Group (UK) Ltd, Croydon, CR0 4YY
27/03/2026
02079975-0004